Simple Electronics for Modellers

Ian R. Sinclair
B.Sc., MIEE, M. Inst. P

Simple Electronics for Modellers

Model and Allied Publications
Argus Books Limited
14 St James Road, Watford
Hertfordshire, England

Model and Allied Publications
Argus Books Limited
14 St James Road, Watford
Hertfordshire, England

First published 1977

Second impression 1980

ISBN 0 85242 515 5

Printed in Great Britain by A. Wheaton & Co. Ltd., Exeter.

CONTENTS

Other books by Ian Sinclair

INTRODUCING AMATEUR ELECTRONICS
UNDERSTANDING ELECTRONIC COMPONENTS
UNDERSTANDING ELECTRONIC CIRCUITS
INTRODUCING ELECTRONIC SYSTEMS
THE OSCILLOSCOPE IN USE
AUDIO AMPLIFIERS FOR THE HOME CONSTRUCTOR

PREFACE

In the days when electronics meant using large heavy valves, the idea of using electronics in modelling seemed quite absurd. The development of the transistor, and, more recently, the integrated circuit, has now made it possible to make electronic circuits of almost unbelievably small size, so that many applications of electronics to modelling are possible and advantageous.

This book sets out not to act as an instruction manual but rather as a source book, not providing cut-and-dried electronic circuits but rather a collection of working circuits and ideas of how they work and how they may be used. Since the requirements for electronics in each modelling activity vary so much a book of designs is of only limited use, and an understanding of what electronic circuits can do is more useful, so that the modeller can adapt circuits to his own needs.

Since the skills needed for modelling are so similar to those needed for electronics construction with modern components, I hope that this book will also awaken some interest in the wider uses of electronics in our friends who use solder to seal tanks, join rails or seam brass. I for my own part have greatly enjoyed writing a book which reminds me of many happy modelling hours.

Of all the uses of electronics, radio control has deliberately been excluded on grounds both of specialisation and of space; there are, in any case, many excellent texts dealing with this fascinating branch of electronics and electromechanical engineering.

I. R. SINCLAIR

CHAPTER ONE

COMPONENTS AND CONSTRUCTION

AN ELECTRONIC CIRCUIT is one type of electric circuit in which electric currents are controlled without the use of moving parts; in addition the currents used for control are very small. Since electronic circuits are electrical circuits, the usual laws of electrical circuits will apply, but we must remember that laws which apply to a.c. are not all the same as those which apply to d.c.

Direct Current

D.C. is an abbreviation of direct current, meaning a current which flows in one direction only and which remains at a steady value unless we choose to change it. Such a current will come from a battery which has a steady voltage, or from a power supply which converts a.c. into d.c. We use d.c. for small motors, relays, solenoids and for supplying the operating voltage to electronic circuits. The important quantities in d.c. circuits are the supply voltage, often called the e.m.f. (electromotive force) when it is measured with no current flowing, and the circuit resistance.

Electric current is a flow of fragments of atoms, called electrons, in a material, and is measured in units of amperes, usually shortened to amps. One amp represents a very large number of electrons passing through the circuit per second, and we use smaller units, the milliamp (mA), which is 1/1000A, and the microamp (μA), which is 1/1000000A. Even the microamp represents 6250000000000 electrons per second passing any point in the circuit.

Just as heat flows from a place which is hot to one which is cold, and never in the opposite direction, so electric current flows from a place at high voltage to one at low voltage. The quantity we call voltage or potential is, like height or temperature, a measure of the ability to cause movement, in this case of electrons. The units are volts, also used to measure e.m.f., and we sometimes use the millivolt (mV) which is 1/1000V and the kilovolt (kV) which is 1000V.

The rules of d.c. are:

(1) Current flows only in a complete circuit.

(2) No current is created or destroyed at any place in the circuit.

(3) The voltages added up round the circuit equal the total e.m.f. from the battery or power supply.

(4) When current passes through a resistor, there is a voltage across the resistor equal to resistance × current.

Current I amps

Resistor, value R ohms

Voltage V volts

$V = R \times I$

or $I = \frac{V}{R}$

or $R = \frac{V}{I}$

Fig. 1.1 Ohm's Law.

Ohm's Law

The last of these rules is called Ohm's Law, and means that in d.c. circuits we can find values of voltage, current or resistance provided that we know two of the values. If, for example, we know the current through and voltage across a resistance, we can find the value of resistance (which is voltage/current), or if we know the value of current through a known value of resistor, we can find what voltage must be across it (equal to resistance x current). The units of resistance are ohms if voltage is measured in volts and current in amps.

Measurements of d.c. voltage and current, when they are needed, must be made using voltmeters and ammeters. For the small voltages used in electronics, a range of 0 to 10V is usually sufficient, and the currents taken are very small, so that current ranges of 0–1mA and 0–10mA are usually sufficient. What is more important is that the instruments are connected properly and that their limitations are known.

Voltmeters are connected **across** the voltage to be measured, with the sign of voltage correct – the meter connected with its + (red) lead to the more positive end of the voltage to be measured. Current meters can be used only if the circuit is broken, and the meter wired so that the current must pass **through** it. Current readings are more reliable, however, because most types of

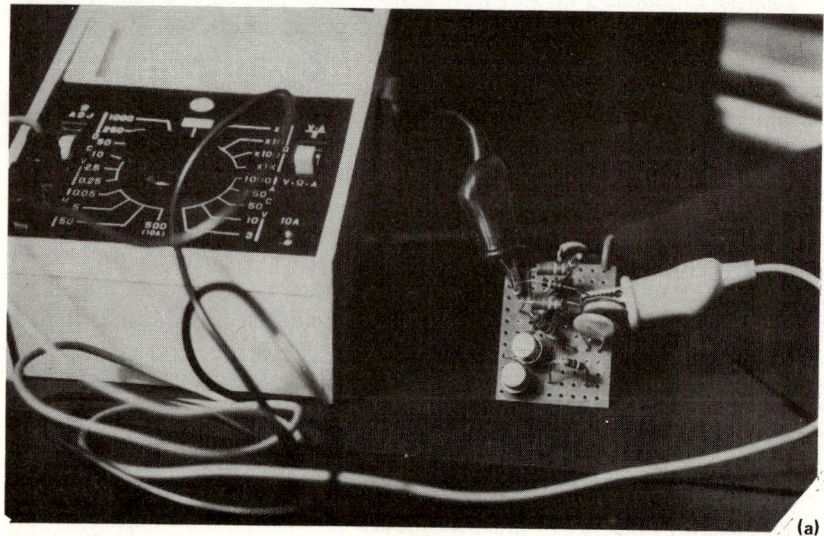

(a)

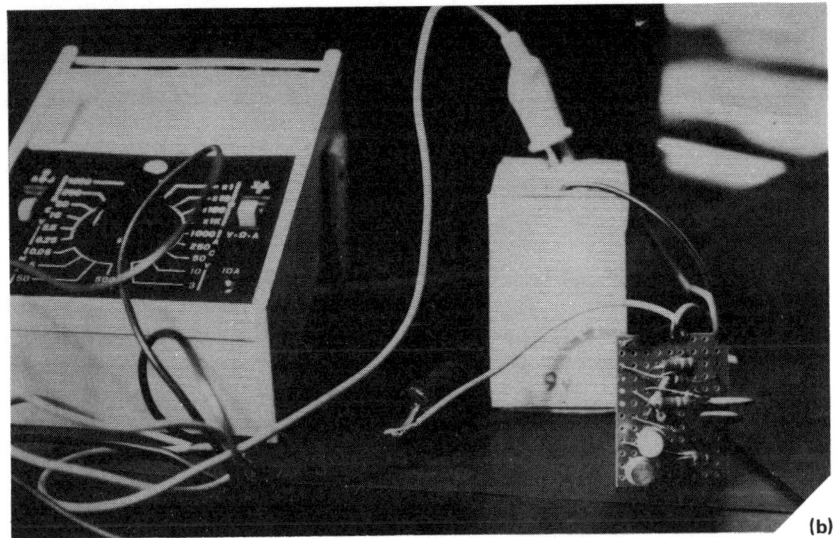

Fig. 1.2 Placing a voltmeter (a) and a milliammeter (b) in circuit. The voltmeter leads are placed one on each side of the component across which the voltage is to be measured; most measurements are made with one voltmeter lead connected to negative (earth). Using the milliammeter involves breaking into the circuit and connecting the milliammeter across the break.

meters which are used are operated by currents, and even when voltage is being measured some current is taken by the meter. This makes the readings of voltages on the low side, unless the meter takes very little current indeed (10μA or less). This point is taken up again in the chapter devoted to measurements.

Alternating Current

Alternating current is, as the name suggests, current which continually changes, going alternately in one direction and then in the opposite direction in the circuit many times per second. Closely related to alternating current are electronic signals which may be alternating currents of various types, or pulse waveforms, in which the voltage rises and falls suddenly.

When we speak of voltage or current in an a.c. circuit we mean something rather different from the steady values we use in d.c. circuits. Our idea of a steady a.c. voltage is one which varies between set limits, the peak voltages. The total voltage from the negative peak to the positive peak is called the peak-to-peak voltage value; similarly we can measure a peak-to-peak current value. Our idea of a steady a.c. voltage or current is one with a steady value of peak-to-peak voltage or current, even though the values of voltage and current are changing all the time and equal to the peak values only for brief times.

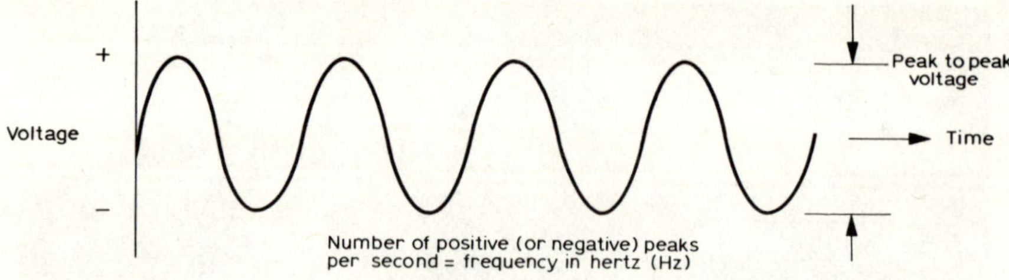

Fig. 1.3 Graph of voltage and time for a.c.

We need more than a measurement of peak-to-peak voltage or current to describe alternating current completely. When we plot a graph of the voltage against a scale of time, the a.c. is represented by a repeating shape, the wave-shape or waveform. The number of repetitions of the waveform per second is a quantity called the frequency and this quantity, along with the geometrical shape of the waveform, is important if we are to make a.c. measurements.

The mains voltage repeats slowly, only 50 times per second (50 Hertz), and its waveshape is the most smoothly-varying known, called the sinewave. When sinewaves are used in circuits, measurements of r.m.s. (root mean square) voltage and current are often used. These are values of the equivalent d.c. voltage, meaning the d.c. voltage which would give the same power output in heaters, motors and similar appliances. The r.m.s. value is given by $\frac{\text{peak-to-peak value}}{2 \cdot 8}$ for a sinewave only. R.M.S. values and meter measurements of a.c. are seldom used in electronics.

Passive Components

Resistors, capacitors, inductors and transformers are called passive components because they cannot add any power to an a.c. signal fed into them. Resistors are small components used for controlling currents and changing the voltage levels of a.c. or d.c. Capacitors come in a wider variety of physical sizes, and will pass a.c. from one part of a circuit to another without allowing d.c. to flow. Transformers perform a similar function, but can also change the voltage of the a.c., up as well as down. Inductors pass d.c. easily, but limit the flow of a.c.

When these passive components are used in combination, we find some other effects. When a resistor takes current to a capacitor, from a steady voltage, the voltage across the capacitor does not rise instantly, but at a rate which depends on the values of capacitance and resistance. This combination is called a time-constant, because it can be used for timing, as we shall see later. The combination of inductor and capacitor makes a tuned circuit (also possible with resistor and capacitor) which passes one frequency in preference to any other.

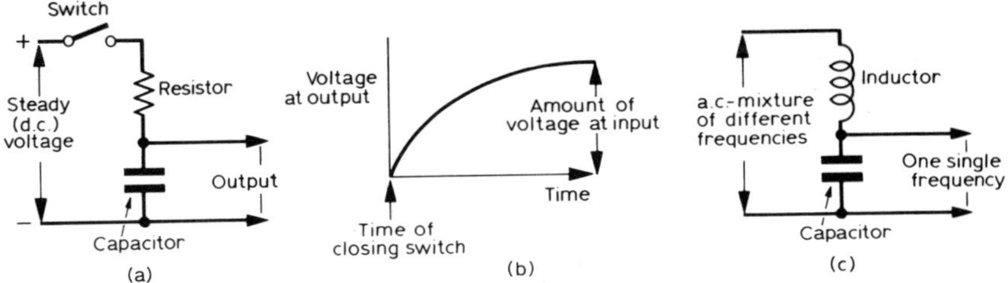

Fig. 1.4 Combining components. (a) resistor and capacitor circuit, (b) the output voltage after switching on, (c) inductor and capacitor.

Active Components

Transistors, thyristors and integrated circuits, all members of the semiconductor family of electronic components, are classed as active components. They need a power supply, usually d.c. of a specified voltage, and they can be used to increase the power of a signal, a.c. or d.c., which is fed to them. Strictly speaking, they are power controllers rather than amplifiers, and the action which we call amplification is really the creation of a new signal of greater power under the control of the signal which is fed to the semiconductor.

Transistors, thyristors, and triacs (see later) are called discrete components because each of them carries out one single task of controlling the output of a voltage, a.c. or d.c. When we wire several components together, we can carry out much more complicated actions, from a comparatively simple amplifier for a record player to the quite incredible complexity of a calculator.

Steady advances in the manufacturing processes for making transistors have resulted in techniques which can make complete circuits containing several transistors and other components, sometimes in thousands, one small piece (chip) of material the same size as would be used for a largish transistor. This is the famous integrated circuit (IC), and we shall be making good use of several types of integrated circuit in this book.

Integrated Circuits

Unlike individual transistors, which can be wired into any type of circuit we design, integrated circuits are designed each to do a specific job. Some are too specialised for the type of work we need; television signal processing circuits, for example, designed to separate the colour signals from the black/white signal. Others can be put to a large variety of uses; timer circuits, operational amplifiers, and logic circuits.

Unlike transistors, which are usually soldered directly into circuit, the ICs can be mounted in holders like old-fashioned valves. This is not because they need to be replaced, but because it is then easier to check the whole circuit thoroughly before the IC is used. Since ICs cannot be unsoldered so easily from a circuit as can transistors, and may cost rather more (though much less than all the transistors they replace), it makes sense to be able to change a circuit or an IC without risking damage.

Excessive heat from a soldering iron is a hazard to any semiconductor, and this applies particularly to ICs, which have thicker mounting pins to conduct heat. In addition, of course, the use of holders makes it possible to experiment and to transfer an IC from one circuit to another improved version.

Circuit Diagrams

A circuit diagram is a way of showing electrical connections. Each component in an electronic circuit will have two or more terminals, usually wires coming out of the component. Each terminal will have to be connected to some point in the circuit, perhaps to another terminal on another component, perhaps to a common earth or positive line. These connections may be made using wire or copper strip; there may be many possible practical layouts, but only the connections on the circuit diagram must be made, for the circuit diagram is a map showing what connections are made to each terminal.

Where two terminals are shown connected, a wire or copper strip must connect them; where several connections are made to one line on the diagram, these terminals must be connected to one wire or strip on the practical layout. A short circuit means than an unintended connection has been made, an open circuit means that some intended connection has not been made. Our practical layout should have each connection shown on the circuit diagram and none which are unintended.

Practical Layouts

For most purposes, the practical layout will be one which is convenient from the point of view of fitting into the available space or having controls which are easy to use. For example, the part circuit diagram of Fig. 1.5a shows a potentiometer VR1 used to control the amount of signal passing from transisfor Tr1 to transistor Tr2. The circuit diagram shows what points are connected; that one end terminal of the potentiometer is connected to one terminal (–) of the capacitor C1, the other end terminal of the potentiometer is connected to earth and the middle (sliding) contact terminal to the (–) terminal of capacitor C2.

This does not mean that these components must be close together, and one practical layout might be that of Fig. 1.5b, with Tr1, Tr2, C1, C2 and the fixed-value resistors on a circuit board and the potentiometer separate, using three wires to connect the potentiometer terminals to the correct points on the board. In such a case we would use coloured insulation on the wires to remind us of the correct connections. In addition to this, the order of components on the board need not be that shown on the circuit diagram; the transistors could be placed next to each other, for example, but the connections must be these shown on the circuit diagram and no others.

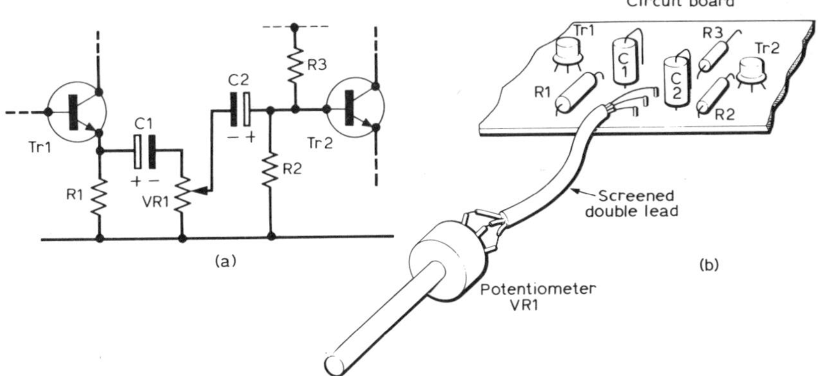

Fig. 1.5 (a) part of circuit diagram, (b) a possible practical layout.

Not all circuits can be made in any form we choose. Circuits using high frequencies, including some parts of radio control transmitters and receivers, must follow strict rules about their practical layout. The circuits in this book are intended to form the basis of practical arrangements which can be varied to suit the uses to which the circuits are put and none of them should be in any way critical about layout provided that long straggling wires are not used to carry signal currents. Some suggested layouts are given, and Chapter 9 deals with the construction of printed circuit boards.

Circuit Boards

Wherever possible, most of an electronic circuit will be built on a circuit board. The type of circuit board most suitable for one-off circuits is the strip-

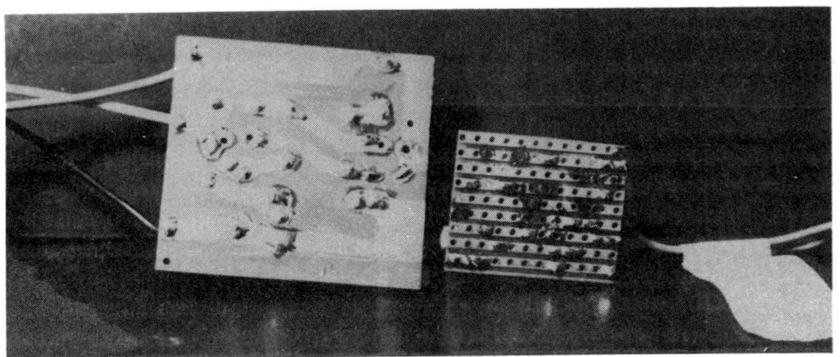

Fig. 1.6 Circuit boards – *Veroboard* (right), PCB (left). In the boards illustrated, the components have been mounted upright on the *Veroboard* so that the board area is smaller.

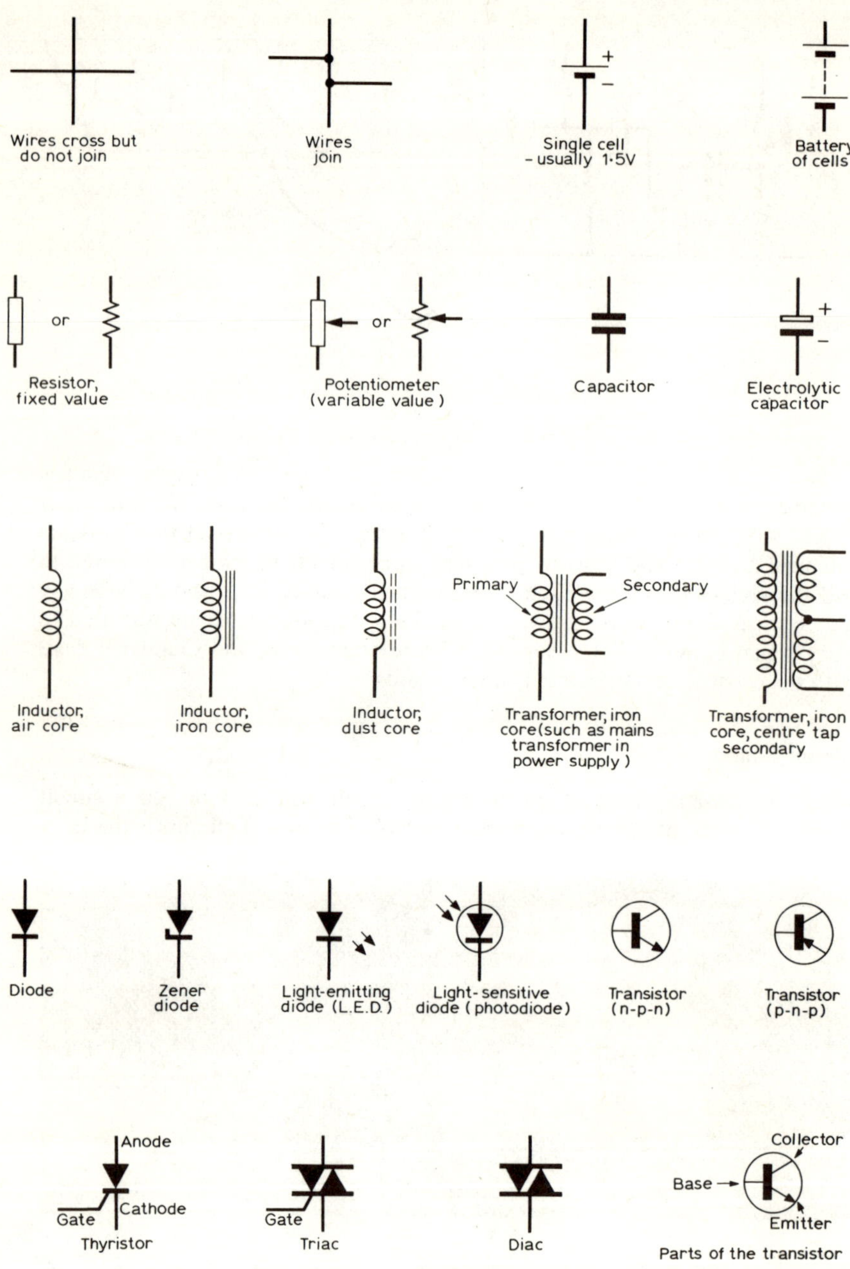

Fig. 1.7 Symbols for circuit diagrams.

board, such as *Veroboard*, which consists of copper strips at spacings of 0·15" or 0·1" according to the type of board used. The strips are drilled with a series of holes spaced at the same 0·1" or 0·15" setting as the strips themselves. The 0·15" board is easier to use, particularly for the beginner to small-scale soldering work, but the 0·1" board makes smaller assemblies possible, and must be used when ICs are used, because 0·1" (2·5mm) is a standard IC pin spacing. When the closer spaced strips are used, care must be taken when components are soldered to the strip, as it is only too easy to spread the solder on to another strip. A fine-tipped iron and small gauge solder helps.

When the practical layout has been determined, using drawings on squared paper, and remembering that the components are mounted on the opposite side of the board from the copper, the components can be soldered in place. The components may be mounted flat against the board, raised off the board, or upright, and the terminal pins protrude through the holes in the copper strip so that they can be soldered to the copper. Take great care in the layout to avoid the common mistake of soldering both pins of a component to the same piece of copper strip. Layouts will be discussed further as each circuit is introduced.

Soldering

The soldering work needed in electronic circuit construction is quite different from the soldering needed for modelling activities. We are not concerned with making seam joints in tinplate or brass, but with making good electrical connections which are also strong enough to stay in place. No acid fluxes, such as the time-honoured Baker's Fluid, should ever be used for electronic or any other electrical work; and we certainly do not have to pickle the components in acid after soldering! The solder used should be resin cored, such as *Multicore*, and a small (15W to 25W) electric iron with a fine bit is needed.

Most of the joints which we make are between metals which are easy to solder and we need only ensure that both metals are clean. Taking a typical example of soldering a resistor in place on a piece of *Veroboard*, the metal strips on the board should be cleaned with fine glass paper if there is any corrosion on them, as should the wire of the resistor. The leadout wires of the resistors are then bent so that they fit into place on the board, which is now clamped to a stand or vice. The resistor is now put into place and held temporarily by clips or any other method available. Fingers are not sufficiently heatproof, and are not recommended.

The clean hot soldering iron is then placed on the joint so that it heats up both the wire of the resistor and the copper strip on the board. The solder is now brought up to the joint and allowed to flow until the wire and the strip are coated with solder, and the hole cannot be seen. This does not require a great blob of solder. When the solder has coated the joint sufficiently and smoke (from the resin) no longer rises from the joint, the iron is taken away, and the soldered joint allowed to cool. The solder must be allowed to set be-

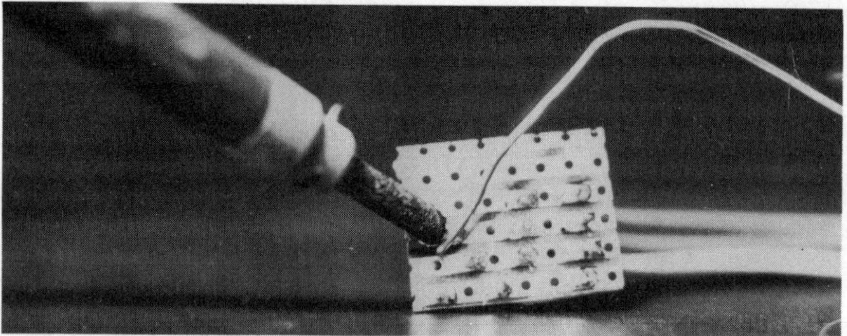

Fig. 1.8 Soldering. The iron is held firmly against the work and the solder brought up to the hot metal.

fore the joint is disturbed, because any movement before the solder has completely set can cause what is called a "dry joint", in which the solder has a crystalline appearance.

Dry joints have variable electrical resistance, and cause unreliable operation of circuits; because they are not easily traced in a complete circuit they must be avoided at all costs. An equally poor joint can be made if the iron is held against the joint for too long, so that the solder oxidizes, the copper strip begins to peel away from the board, and the board itself starts to scorch.

A properly soldered joint should be good both mechanically and electrically. If the solder refuses to spread, the metal surfaces are probably dirty and the remedy is simple. One of the commonest faults of beginners to soldering is to dab the iron against the work, never quite melting the solder properly. If you have once seen the solder flow properly on to a joint, you are never again likely to make the mistake of dabbing.

Difficult Joints

A few metals are difficult to solder, though the most difficult are not used for terminal wires. Nichrome wire, used for electric fire elements is one, but one we are not likely to use. Another difficult nickel alloy, however, is the one used for sealing into glass, which is found on transistors, diodes and other semiconductors. Such wires are often gold-plated to make soldering easier, but the unplated wires which are still common, particularly on bargain pack transistors, are not easy to solder until the correct technique is used.

The correct method is to 'tin' the wire before soldering it into place. Tinning means putting a thin coating of solder on the wire before trying to solder it on to the board, and the reason for doing it is that the nickel-alloy needs a rather higher temperature than copper for successful soldering. This can be more easily done when the wire is soldered by itself, with no copper strips to conduct heat away from the soldering iron.

The wire is carefully cleaned, taking care not to pull it away from the body of the transistor or diode, nor to bend it sharply near the body, since the seal between the wire and the glass which is used as an insulator must not be broken. The easiest method is to hold the wire in a pair of small pliers and pull folded glass-paper over the exposed part of the wire. Now, still using the pliers to hold the wire, touch the end of the wire to the iron and bring up the solder. As the solder melts, rub the tip of the iron over the wire until a film of solder forms on the end of the wire. Remove any blobs and allow the wire to cool before releasing it from the pliers.

The pliers, as well as protecting the transistor from mechanical damage, also prevent heat from the iron travelling up the wire to damage the semiconductor material. All three leads of a transistor will have to be tinned in this way, and when the transistor is soldered into the circuit board it is preferable (though not completely essential, because the joint can be made quickly on tinned wire) to use pliers in the same way to avoid heat damage.

Desoldering

Sometimes solder has to be removed, either because a component has been wrongly placed or because too much solder has been put on a joint. Desoldering tools can be bought, but the expense is not justified unless a lot of desoldering has to be done as a matter of course. The simple method of getting rid of blobs of solder is to heat up the iron, shake or wipe it free of excess solder, and apply it for a short time to the faulty joint. Some solder will then flow from the joint on to the iron. If this is then repeated, solder will be removed each time, and the joint can be completely stripped of all but a thin film of solder so that the component can be removed if needed.

This simple method works well for cleaning up a joint or for removing a resistor or capacitor, but is not so suitable for removing semiconductors, particularly ICs, or their holders. Many ICs have fourteen or more pins, and it is impossible to remove enough solder to take all of the pins out together. When an IC is known to be faulty, it is usually removed by cutting the pins, which can then be taken out one by one. Our problem is much more likely to be that the IC was incorrectly placed in circuit, since we do not have the test-gear to check for a faulty IC. For this reason, it is much better, as advised earlier, to use holders for ICs, and to solder the holders in place.

ICs can be plugged in when the circuit is checked out, and the voltages on each of the pins can be checked before the IC is put at risk. If a wiring mistake has been made, the holder will have to be removed for resoldering. This can be done by clipping the pins of the holder, but is undesirable, because the holders are often more costly than the ICs (they cannot be made by a completely automatic method), and the use of desoldering braid is preferable.

Desoldering braid looks like thin lampwick made of copper wire, and the action is the same. When the braid is dipped into hot solder, the capillary wick action causes the solder to run along the braid in preference to any other action, so that solder can be removed from a joint. To use desoldering braid, the end of the braid is placed directly over the joint which is to be desoldered; if needed, the braid can be wrapped for one turn round the joint. The hot iron is then laid against the braid, so that the heat is conducted through the braid into the joint, and the braid is always hotter than the joint.

As the solder melts, it runs into the braid and is absorbed. Both braid and iron are then removed, leaving the joint free of all but the thinnest coating of solder, and with no solder joining the component lead-out wire to the circuit board. When the braid cools, the end which is stiff with solder can be cut off and discarded, so that a fresh piece of braid is ready for use.

CHAPTER TWO

CURRENT AND VOLTAGE CONTROL

MANY MODELLING ACTIVITIES require the use of small relays and, more particularly, solenoids. These require definite current and voltage levels to operate with certainty, and also need some method of control; usually that of switching on and off. This action may, however, have to be done from a remote switch through thin wire or uncertain contacts, or may be done automatically as will be shown later. This chapter is devoted to the control of solenoid-like devices using much lower control currents by the use of electronics.

Action of Solenoids

Solenoids consist of a coil of wire wound round a core of "soft" magnetic material. Soft in this context means that the material will be strongly magnetised when a steady (d.c.) current flows in the coil of wire, but demagnetised when the current ceases to flow. Sliding into a cylindrical hole in the core is a steel plunger which can be attached to an arm and whose movement can be used to actuate any other mechanical mechanism. When steady current flows in the solenoid coil the plunger is attracted into the core; when the current is switched off a spring returns the plunger to its previous position.

The disadvantage of the straightforward solenoid of this type for modelling purposes is that current is being consumed all the time the solenoid is held in. This can be overcome in two ways. One way is to polarise the plunger rod, magnetise it so that it can be repelled as well as attracted. In this way, one direction of current in the solenoid will pull the plunger in, the other direction will push it out.

Latching solenoids perform the same action by a mechanical method. In a latching solenoid, two solenoid coils are used, one to pull the plunger in and latch it in place, using a mechanical clip to hold the plunger in. The other coil releases the clip, unlatching the solenoid so that the spring can return it.

In applications such as model railway points systems, the simple solenoid can be used, because the power consumed can be supplied from a mains-operated supply. In model aircraft, boats and cars which use the solenoids as part of a radio control system, batteries must be conserved, and actuator systems tend to use as little electric power as possible.

Operating the Solenoid

The traditional way of operating the solenoid which is not part of a radio control system is simply to switch the current on and off. Electronic controls offer the possibility of switching much smaller currents, of using a.c. instead of d.c. for switching, of controlling solenoids from light or temperature levels and simpler switching of latching or polarised solenoids.

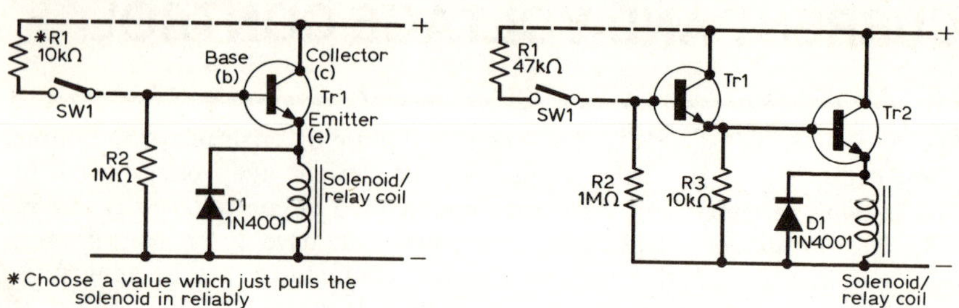

Fig. 2.1 (Left). Solenoid/relay control circuit. The transistor can be a general purpose type such as the 2N697 or 2N3053 unless the solenoid needs a considerable operating current. The voltage used will be the normal relay operating voltage, usually 12V.

Fig. 2.2 (Right). A more sensitive unit than that of Fig. 2.1. Note, however, that the voltage across the solenoid will be at least 1V less than the supply voltage.

Figure 2.1 shows a solenoid control circuit using one transistor. When the switch Sw1 is open, the base (b) connection of the transistor is connected to the supply negative (earth) through the resistor R2. With the base of a transistor connected this way, no current will flow through the transistor; it acts like an open circuit so that the solenoid remains unenergised even though the battery or power supply is connected and voltage is applied to the other terminals of the transistor.

When the switch Sw1 is closed, a very small current flows through the resistor R1 into R2 and into the base terminal of the transistor. If the amount of current passing is enough to raise the voltage across R to about 0·5V, the base terminal will take some current (measured in millionths of amps), and current will flow between the other two terminals. The current flowing between the other two terminals is very much greater than the current flowing into the base terminal, so that the transistor acts as a current amplifier. The current flowing between the other two terminals (emitter and collector) will also flow through the solenoid, and if enough current flows, causing the correct operating voltage to appear across the solenoid, the solenoid will switch on.

In this particular circuit, the current in the solenoid will be sufficient to switch it on only if (*a*) the voltage at the base of the transistor is about 0·5V higher than the voltage needed by the solenoid and (*b*) the current into the base is about 1/100 or so of the current needed by the solenoid. When Sw1 is switched off again, the current through the transistor switches off equally quickly. This causes the voltage at the emitter of the transistor, connected to the solenoid coil, to change sharply because when the current through any coil of wire is interrupted there is a sharp change of voltage.

Left to itself, this would cause the voltage at the emitter of the transistor to become momentarily negative and large, possibly damaging the transistor, so that the diode D1 is added as a protection to by-pass such current to earth. The diode is connected so that in normal operation it is non-conducting.

Low Current

The advantage of this circuit is that the amount of current handled by Sw1 is very small, probably less than a milliamp (1mA = 1/1000A) if the solenoid is a small one. In a d.c. railway system, for example, all the solenoids could be powered from a single supply line, and the control currents passed by very thin control wires. Separating power from control in this way makes it much easier to add to a system, since only the thin control wire has to be added. Control switches can be much smaller than would be the case if they had to operate with the solenoid current through them; if more sensitive circuits are used, the switch can consist of two minute contacts, and the touch of the finger across the contacts will operate the solenoid.

For much more sensitive controls, a second transistor can be added, as shown in Fig 2.2. This makes the current needed for control a few microamps (μA = 1/1000000A), but the voltage needed to switch on at the base of the first transistor is now 1V above the solenoid operating voltage.

Building a Solenoid Control Unit

The same techniques can be used for building all electronic circuits when all of the main components are soldered on to circuit boards. Assuming that for such simple circuits we are going to use *Veroboard* or similar strip-board rather than specially etched-copper boards (see Chapter 9), the procedure is as follows.

The circuit is traced out and a letter marked on each independent point in the circuit. By an independent point, we mean each part of the circuit where one component is joined to others, but which is separated from other connections. Fig. 2.3 shows the two circuits we have shown lettered up in this way. Each letter now represents a strip of copper in the circuit board. We now take a piece of circuit board with at least as many strips as we have letters, and mark on it the letters we have used to mark the circuit, in order, one letter to each strip.

Do not confuse the letters on the strips with the e, b, c, letters used to identify the emitter, base and collector of the transistor; it is better to remember the symbols for these terminals and their placing on the transistor than to have to use any other identification. With both the circuit diagram and the board lettered, it is now easy to connect up the circuit components.

Looking at the circuit and the board for Fig. 2.1, we can see that the

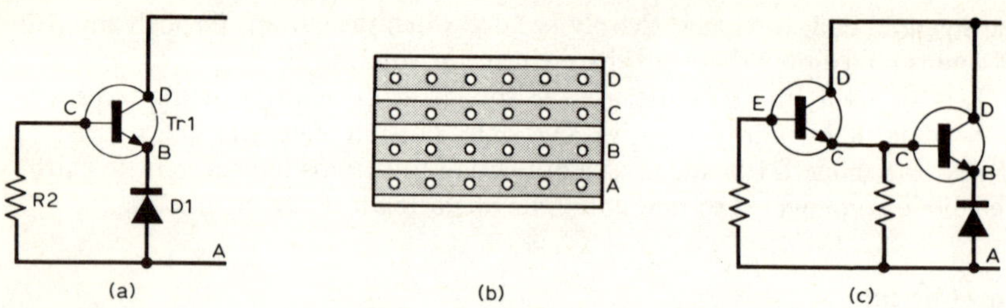

Fig. 2.3 Lettering up the circuits for *Veroboard*. Draw out the circuit diagram and letter the independent points (see text). Try to make the letters used for the transistor connections consecutive. (a) circuit of Fig. 2.1., (b) appearance of underside of the strip-board, (c) lettering for the circuit of Fig. 2.2.

transistor is connected to the board with its emitter soldered to strip B, its base to strip C and its collector to strip D. The leadout wires of the transistor should be tinned as advised in the previous chapter, remembering to hold them with pliers near the point where they are sealed into the body of the transistor.

The leads must then be identified. For transistors such as the 2N697, which is particularly cheap and convenient for experimental and general-purpose use, the pattern of the leadout wires is a triangle, arranged as in Fig. 2.4, in the sequence emitter-base-collector. On this particular transistor, the collector connection is also made to the metal case of the transistor, which should not touch against any other circuit component. This sequence of lead-out wires is very common on small transistors, but for reasons known only to the manufacturers, some types use quite different sequences and arrangements. We shall include the connections diagram for all the transistors mentioned in this book.

The tinned leads are now inserted through the holes in the circuit board strips, with the body of the transistor on the fibre side of the board and the ends of the leads protruding through the holes at the copper strip side. Check that the connections you are about to make are correct, and then solder the ends of the leads to their copper strips. The transistor may be soldered to the ends of the strips or a few holes in, it makes no difference to the connections which are being made. With experience, very compact arrangements are possible.

The diode can now be connected between strips A and B, and it is essential to connect it the correct way round. The connection which is symbolised by the line (the cathode) connects to strip B; the connection symbolised by the arrowhead (the anode) connects to strip A. On a diode, the cathode end may be marked red or, rather confusingly, with a + sign. A few diodes may have no markings, and the correct polarity must be found by experiment.

The circuit of Fig. 2.5 will help to find the polarity of any diode. The cell is a single 1·5V torch cell and the resistor is a 1·5KΩ of any wattage. The meter is a 1mA meter or indicator. With the diode connected one way round, no current (or very little) will be indicated by the meter. With the reverse connec-

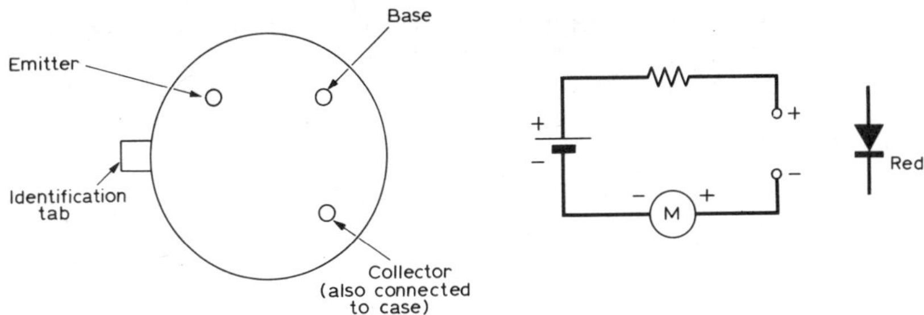

Fig. 2.4 (Left). Connections for transistors type 2N697 or 2N3053, viewed from the underside.

Fig. 2.5 (Right). Testing the polarity of an unmarked diode. When a diode does not conduct, mark the end connected to the + terminal with a blob of red paint.

tion, much greater current will be indicated. Try the diode each way round. When the lower or zero current is indicated, the cathode end of the diode is connected to the + pole of the battery and the diode can be so marked with quick-drying red paint.

The resistor R2 can now be connected between strips C and A, and soldered in place. Unlike transistors and diodes, resistors can be connected either way round.

Flexible Connections

For most modelling applications of this circuit, no more components will be connected on to the board, which need be only a few holes long, because both the solenoid L1 and the switching circuit R, Sw1 will be elsewhere. For example, in a model railway points switching application the solenoid would be located at the trackside, the circuitboard under the baseboard or in a dummy toolbox at the trackside, and the switch and resistor R1 at the control panel.

We must leave room on the circuit board for more connections to strips A, B, C and D to lead flexible connections to these remote parts of the circuit. Strip A is led to the negative terminal of the power supply and also to the negative terminal of the solenoid. Strip B is connected to the other terminal of the solenoid. Strip C is taken, through as fine wire as is needed, to one terminal of Sw1, the points operating switch. At the switch end of the circuit, the other terminal of the switch is connected to one end of R1, and the other end of R1 is connected to the positive end of the power supply. Strip D is also connected to the positive end of the power supply.

The circuit of Fig. 2.2 can be similarly built on a circuit board, but the board will need to have five strips, seven holes in length, and a much larger value of resistor, R1, in series with the switch Sw1. Note, by the way, that there is no obvious connection between the amount of resistance in ohms, kΩ or MΩ and the physical size of a resistor. The physical size is proportional to the power which can be dissipated, so that for example a 1/2W resistor is larger than an 1/8W.

Latching Solenoids

Latching solenoids need a short burst of current to lock on, and another short burst of current to another coil to turn them off; with conventional connections this usually is done by pushbutton switching, and there is no indication of which button was previously pressed except by looking at the solenoid. Electronic operation of latching relays makes it possible to operate with a normal on/off switch, so indicating whether the solenoid is on or off, and with very little current flowing so that very little power is consumed.

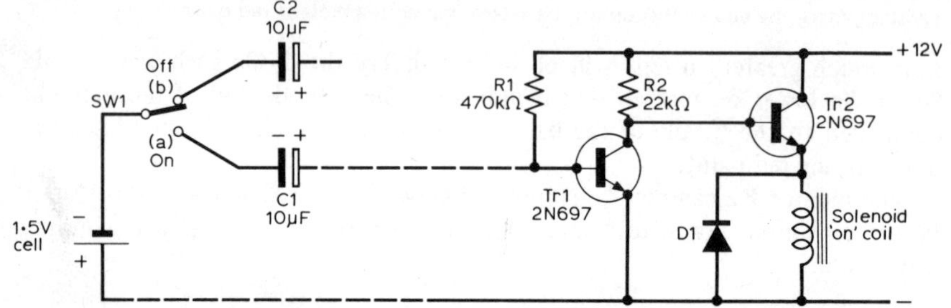

Fig. 2.6 Operating a latching solenoid or relay. The part of the circuit containing the transistors needs to be duplicated for each coil of the solenoid or relay.

The circuit used for each coil of the solenoid is shown in Fig. 2.6, where Tr2 is a transistor which supplies current to a coil of the latching solenoid using the same circuit as was used in Fig. 2.1. Tr1 is another transistor which is connected as a voltage amplifier, so that a small change of voltage at the base of Tr1 will cause a large change of voltage at the collector of Tr1. The base of Tr1 is connected to the positive supply voltage through the resistor R1, whose value of resistance is high.

This connection ensures that in normal operation, current flows into the base of Tr1, keeping the transistor switched on, so that Tr1 behaves as if it had a low value of resistance between its emitter and its collector. This in turn means that the voltage between the collector and the emitter is low, too low to pass current into the base of Tr2. As a result, Tr2 remains off and no current flows in the coil of the solenoid.

When the operating switch Sw1 is turned to position a, the capacitor C1 is connected to the negative pole of the 1·5V cell, so that the voltage on the plate of the capacitor which is connected to the switch suddenly drops by 1·5V. The action of a capacitor is to transmit sudden changes of voltage, so that the other plate also becomes 1·5V negative. This plate is connected to the base of the transistor, and the change of voltage is enough to stop current flowing in the base of Tr1. With no current flowing in the base of Tr1, no current flows between its collector and emitter either, so that the voltage at the collector rises, and current starts to flow into the base of Tr2, which now switches on, passing current into the solenoid coil and latching it in.

Meanwhile, back at the base of Tr1, current is still flowing in R1, because of the voltage difference between the ends of the resistor, but the current is pouring into the capacitor, raising the voltage on the plate which is connected to the base of Tr1. The voltage at the other plate stays unchanged, because it is still connected to the battery. This process of charging takes time; how much time depends on the values of C1 and R1 which we have chosen. As the voltage rises, it soon becomes high enough to pass current into the base of Tr1 again, switching Tr1 on and restoring the original conditions, so that current stops flowing in Tr2 and in the solenoid coil. The solenoid has now latched, however, and will stay latched until the latch is released by the action of the other coil.

When the switch Sw1 is changed over to position b, the other part of the circuit, Tr3, Tr4, (not shown) acts in the same way to energise the unlatching coil for a short time, so that the solenoid returns to its original position. With the values of components shown, the times for latching and unlatching are about 0·5 second, assuming a smooth 12V supply, which should be time enough for even the slowest of solenoids.

If this time is too long, it can be reduced preferably by using smaller values for C1 and C2. The time will be roughly proportional to the values of C1 and C2 used, so that using half the values shown will result in a switching time of half the original value. Current passes in the solenoid coils only for the time set by the time constant C1R1 and C2R2; when there is no changeover taking place current flows only in Tr1 and Tr3, and this amount of current can be very small, less than 1mA in the circuits shown.

Operation by Light or Temperature

The use of transistors for switching current into solenoids or relays makes it possible to operate these devices from light levels or temperature. All we need to do is to make the light or the temperature cause enough change in voltage at the base of a transistor, which in turn operates the solenoid or relay. Fig. 2.7 shows such a circuit used to turn a relay or solenoid on when the light level of the cell exceeds a preset value.

This circuit uses a cadmium sulphide cell, type ORP12, which is sensitive, though rather slower in operation than some of the photocells we shall use later. The action is that the resistance of the cell varies according to the amount of light falling on it. The approximate resistance levels are 100kΩ in moonlight, 500Ω under bright fluorescent lighting and around 10Ω in sunlight.

In the circuit of Fig. 2.7, the amount of current flowing in R1 and VR1 depends on the resistance of the ORP12. With the cell in the dark, the amount

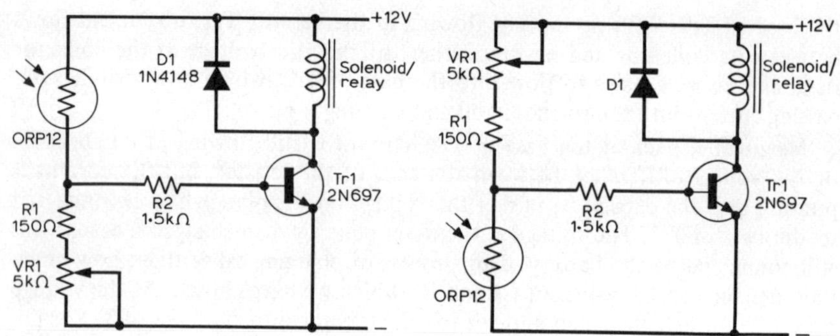

Fig. 2.7 (Left). Operating a solenoid or relay from a light beam. The solenoid is ON when the light level exceeds the level set by the adjustment.

Fig. 2.8 (Right). Modification of the light beam circuit which will turn the solenoid OFF when the light exceeds the set level. (Courtesy RS Components Ltd).

of current will be very small, less than 0·1mA, so that the voltage across R1 and VR1 will be very low, less than 0·5V. This is not enough to turn on the base of Tr1, so that no current flows between the collector and emitter, and the relay or solenoid remains off.

When light strikes the cell, the resistance of the cell drops, more current flows through R1 and VR1 and the voltage at the base of the transistor rises to a value high enough to start current flowing. The current in the collector circuit then turns on the relay or solenoid, which remains on as long as the cell is illuminated. When the light is cut off, the current will cut off about 1/3 of a second later. The delay is caused by the time taken by the cell to stop conducting.

The sensitivity of the circuit is set by VR1, so that a change in light level, which need not be from dark to light but could also be from light to brighter light, can be set to activate the circuit. With the slider of the control VR1 near the negative line, the circuit is most sensitive; the least sensitive position is when the slider is at the 150Ω end of the control.

This circuit can be slightly modified so as to switch on when the light level drops. The cricuit of Fig. 2.8 is used; the only change is that the positions of the cell and the control resistors have been interchanged. VR1 once again sets the sensitivity of the circuit.

Increasing Sensitivity

Very much greater sensitivity is possible if another stage of amplification is used. For greater amplification, it is more convenient to use an integrated circuit than another transistor, though a separate transistor will still have to be used if the solenoid or relay which is being operated takes more than 30mA

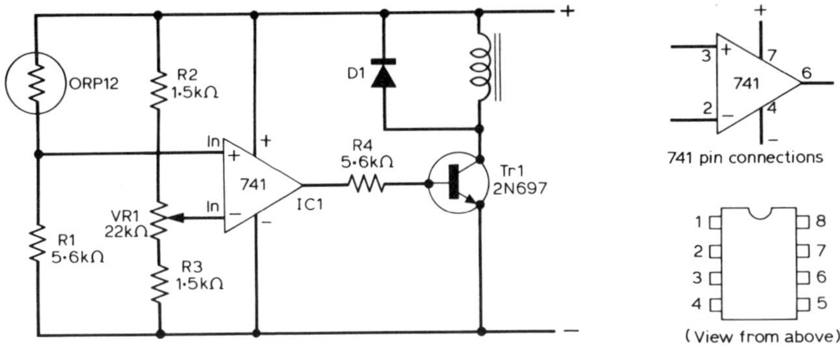

Fig. 2.9 Using an op-amp for greater sensitivity. Note that the op-amp pins are numbered looking from *above*, unlike transistors. (Circuit, RS Components Ltd).

current. The drawing of Fig. 2.9 shows a type 741 operational amplifier used for this purpose. The voltage at the − input of the 741 is set by the potentiometer VR1, and the + input of the 741 is connected to the junction of the photocell and the 5·6kΩ resistor.

The action of the 741 is that it amplifies the difference between the inputs. If the voltage on the + input is greater than the voltage on the − input, then the output voltage is high; if the voltage at the − input is higher than the voltage at the + input, then the voltage is low. The change in voltage between the inputs which is needed to swing the output voltage from high to low or low to high is very small, so that the 741 is a very sensitive detector indeed.

In our circuit, the control VR1 sets the voltage at the − input, and the voltage at the + input is controlled by the photocell. With no light on the photocell, the voltage at the + input is low, and the output voltage is also low, so that Tr1 is off and the relay is off. When the light level on the photocell is high enough, the voltage at the + input just exceeds the voltage at the − input, and the 741 output goes high, switching the transistor on and the relay on.

Hysteresis Action

One possible disadvantage of the previous circuits is that they have the same action in each direction. If the light level is close to the amount needed for switch on, then a very small change in light level will cause the circuit to switch on, and unnoticeable fluctuations in light level can cause the circuit to switch on and off. It is often better to have a circuit which will switch on at one level of light and then remain on until a lower level is reached again. This type of action is called a hysteresis action.

We can introduce a hysteresis action very easily into the circuit when a 741 operational amplifier is used, using the circuit of Fig. 2.10. In this circuit, the + input to the 741 comes from the output voltage rather than from a potentio-

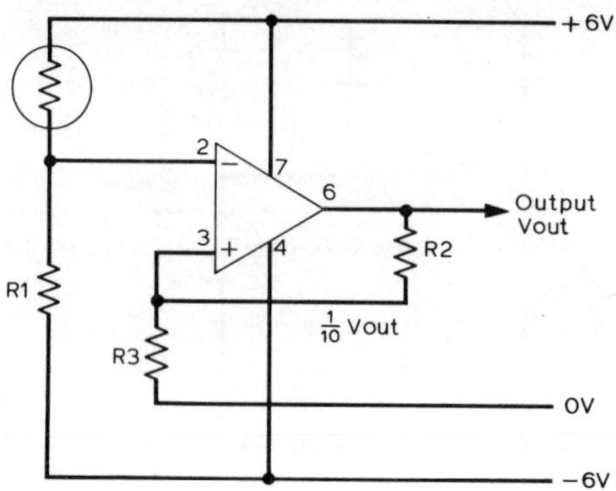

Fig. 2.10 Using a type 741 op-amp to provide hysteresis.

meter control. In this way, the voltage at the + input is always a fixed fraction of the output voltage; typically this might be set to about 1/10. If we imagine the photocell in darkness, then with a low voltage at the - input, the output voltage will be high, + 6 V if the + supply is 6V, and the voltage at the + input will be 1/10 of the output voltage, 0·6V in this example. When the action of light on the photocell causes the voltage at the - input to rise to 0·6V, the 741 will switch over, making the output voltage - 6V. The voltage at the + input will be 1/10 of this, −0·6V, so that the voltage at the - input will have to drop to this new value of −0·6V before the circuit can switch back.

Because the + input has been used to set the hysteresis action, it cannot be used to set the sensitivity, which now must be controlled by altering the value of the resistor in series with the photocell. In addition, this circuit needs volt-

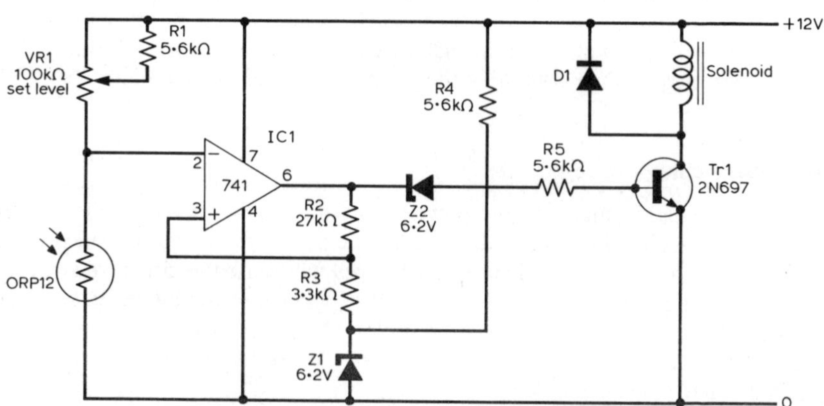

Fig. 2.11 A practical circuit to introduce hysteresis action. Diodes D1 and D2 can be type 1N4148.

ages above and below zero. We can provide operation from a single supply voltage at the cost of some extra components, and the circuit of Fig. 2.11 shows how. A 12V supply is suggested, with the "zero" provided by tapping off a 6·2V supply from a zener diode.

The sensitivity control is provided by a variable resistor in series with the photocell. Because the output voltage of the 741 will be either zero or +12V, we can connect direct to the base of the transistor, but the arrangement of a diode and a resistor shown is better, since it ensures that only a definite switchover from the 741 will operate the relay, and also that excessive current is not drawn by the transistor from the 741.

Calculating Hysteresis

We may not want the hysteresis to be of the amount in the example, so a method of calculating the amount is useful. The hysteresis is set by R2 and R3 in Fig. 2.10, and depends also on the supply voltages used. Suppose the supply voltages used are V+ and V−, then the hysteresis voltage levels will be

$V+ \left(\frac{R3}{R2 + R3}\right)$ for switch-on and $V- \left(\frac{R3}{R2 + R3}\right)$ for switch off.

If we use the single supply voltage system of Fig. 2.11, then these voltages will be found in the same way, but V+ will be the supply voltage positive, less the voltage of zener diode used to fix the mid-voltage, and V− will be the zener voltage. In addition, the hysteresis voltage levels will be added to or subtracted from the zener voltage.

For example, if we used a 12V supply with a 7·5V zener and resistor values of 10kΩ for R2 and 2·2kΩ for R3, then the value of $\frac{R3}{R2 + R3}$ becomes 0·18. The value of V+ is 12V − 7·5V = 4·5V and V− is 7·5V, so that the hysteresis voltages are 4·5 x 0·18V = 0·81V above 7·5V and 7·5 x 0·18 = 1·35V below 7·5. This gives values of 8·31V and 6·15V, so that the circuit will switch over at 8·31V and back again at 6·15V. In general, the greater the ratio of R2 to R3 the smaller the difference in the switching voltages.

All of these circuits can be made to switch in the other direction, that is off, when the light intensity is high, by reversing the positions of the photocell and the resistor in the circuit so that the voltage into the amplifier drops as the light intensity becomes greater.

Temperature Control

We can also operate solenoids and relays from changes in temperature using the same circuits. If a thermistor is used in place of a photocell, the switching action will take place when the temperature is at a point when the resistance of the thermistor is low enough to make the voltage at the input of the amplifier reach its switch-on value. The resistance of a thermistor changes greatly as temperature changes; unlike ordinary resistors, the thermistor resistance becomes lower as its temperature rises. Different thermistors can be selected to cover different ranges of temperature, and also to cover different ranges of resistance value.

Some care is needed in selection, because passing current through a thermistor will raise the temperature of the thermistor, and cause incorrect operation if the current is too high. For experimental purposes, the type TH-2A by RS components (equivalent to CZ1 or VA1005) is ideal; its resistance at 25°C is about 3kΩ, and its resistance will not be noticeably altered by passing currents of a few milliamps. For serious temperature measurement, glass-encapsulated types such as the TH-B18 (or R53) are used, but these are much more expensive.

Operating from A.C. Signals

When we use transistor-operated solenoids, we are no longer confined to controlling the switching with d.c., since transistors will respond to a.c. signals at low or high frequencies. These a.c. frequencies may be electrically generated, or can come from sound waves, ultrasonic waves, or from the high-frequency a.c. waves we call radio waves. When we use wires to carry signals, a.c. and d.c. can share the same wires, because the two can be simply separated. We can, for example, send an a.c. control signal down the same rails as a d.c. train supply. Sound signals from any source can be picked up by a microphone and the electrical a.c. signals from the microphone amplified to provide a control signal.

Leaving aside the business of generating high frequency a.c. signals and amplifying microphone outputs for the moment, Fig. 2.12 shows how a.c. signals can be used to operate a solenoid. The signals must first be separated from any d.c. by using a capacitor. If the capacitance value is small, then low frequency a.c. signals such as a.c. mains frequency will also be filtered out. This is particularly desirable in model railway layouts which are mains-operated, since there is a large amount of a.c. left in the supply to the rails.

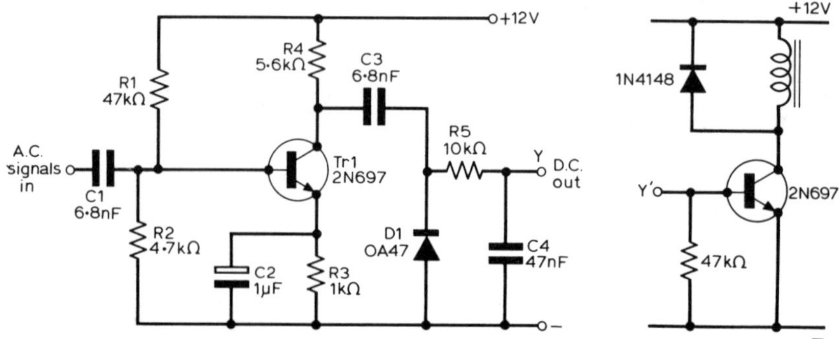

Fig. 2.12 (Left). Converter for a.c. to d.c. signals. An 8-strip circuit board is needed.
Fig. 2.13 (Right). Sensitive switching circuit for solenoid. Y' is connected to the Y output of the circuit of Fig. 2.12.

The transistor Tr1 amplifies the high frequency signal (about 5kHz is suggested) and the amplified signal at the collector of Tr1 is passed through the capacitor C3 to the diode D1 which rectifies the signal, giving a d.c. output at point Y. This output will be sufficient to operate the transistor used to drive a solenoid in the circuits given earlier in this chapter. If the input signal is too small, a more sensitive solenoid operating circuit may be needed, such as that of Fig. 2.13. In this circuit, the voltage at Y′ need be only about 0·5V for operation of the solenoid, and the current into the base of the transistor is only about 1/100 of the current needed to operate the solenoid.

This more sensitive circuit was not shown in earlier applications because it is so easily turned on by small voltage signals, such as could be picked up on long wires or from the rail supply as mentioned earlier. In the application shown here, the filtering effect of the capacitors C1 and C4 ensures that the solenoid will not be operated by stray pulses.

CHAPTER THREE

GENERATING SIGNALS AND DELAYS

WE HAVE SEEN BRIEFLY at the end of Chapter 2 how a.c. signals can be used to operate solenoids or relays. Since we can carry out a large number of other control operations using various types of a.c. waves, some of the circuits which are used to generate these waves must be examined.

For our purposes, a.c. signals mean any repeating waveform whatever its shape. The usefulness of these signals for modelling purposes lies in the fact that they can be separated from d.c. and also from each other, making it possible to control several operations along one signal path, whether it be light beam, wire, sound wave or radio wave.

The Squarewave

The simplest waveform to generate is the squarewave, because one side of a squarewave is generated whenever a circuit is switched on or off. We can generate squarewaves of any amplitude or frequency which we may need, and we can also make use of the quantity called mark-to-space ratio or duty cycle. Referring to Fig. 3.1, the mark time of the squarewave is the ON (duty) time, the fraction of the whole time which is spent at the high voltage level in any one cycle; the space time is the OFF time, the time fraction spent at the low voltage level. A wave of 1:1 mark-to-space ratio (50% duty cycle) will spend half of its time at the high voltage level and the other half at the low voltage

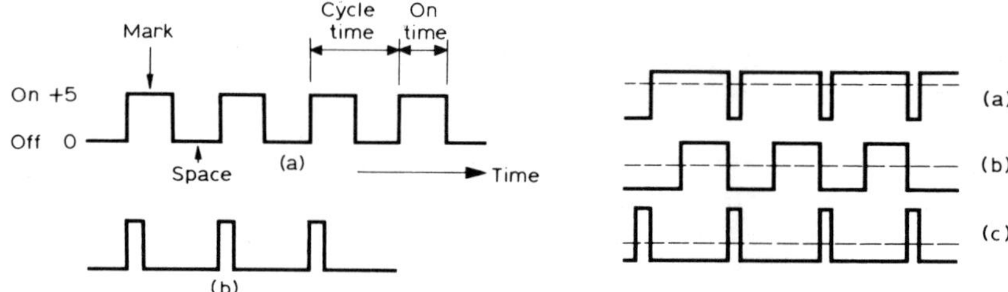

Fig. 3.1 (Left). (a) mark and space. The mark/space ratio here is 1:1, so that the duty cycle is 50%. (b) a 1:9 mark/space ratio, duty cycle 10%.

Fig. 3.2 (Right). Effect of duty cycle on average voltage (shown dotted). (a) 90% duty cycle, (b) 50% duty cycle, (c) 10% duty cycle.

level. A wave of 10% duty cycle has a mark to space ratio of 1:9, with 10% of its time at high voltage and 90% at low voltage.

We can vary the value of the mark-to-space ratio of a wave while the wave is being generated, and we can use this change of mark-to-space ratio to carry information. We can also use changes in mark-to-space ratio to change the amount of energy which is being carried by a signal. For example, if a square wave, whose mark-to-space ratio is being changed is used as the supply to a small motor, then the speed of the motor will change as the mark-to-space ratio is changed.

With a 1:1 mark-to-space ratio, for example, the motor will have running voltage applied to it for half the time of a wave. If the wave were at a low frequency this would cause the motor to run in an irregular way, accelerating during the mark time and slowing down during the space, but if the wave frequency is high, so that the mark and space times are short compared to the time the motor needs to change speed noticeably, the motor speed will be steady.

Because the amplitude of the squarewave can be the normal running voltage of the motor, the torque (turning effort) of the motor will be greater than it would be if we regulated the speed by changing the voltage of the supply to the motor, so that the motor is less likely to stall at low speeds when this type of control system is used. This makes for more realistic speed control.

A variable mark-to-space ratio can also be used to operate solenoids, so that the stroke of the solenoid can be varied rather than switching from off to on; the brightness of lamps can also be varied if they are supplied from a wave with variable mark-to-space ratio. In each case, the times of the marks and spaces are very small compared to the times which the controlled devices take to respond so that there is, for example, no vibration in the solenoid and no visible flicker in the lamp.

Generation of Squarewaves

Electronic generation of squarewaves makes the generation of waves of variable mark-space ratio comparatively simple. In addition, we can generate waves of low amplitude for transmitting over wires or by any other method; these waves can then be amplified with no change in their mark-space ratio. In the past, even electronic methods of generating variable mark-space ratio waves were rather cumbersome, but the availability of cheap integrated circuits has now made the circuitry much simpler. The circuit which we shall use for most of the methods shown in this chapter is the type 555 timer, an IC which is available from several suppliers at very low prices at the time of writing.

Fig. 3.3 shows a circuit which uses the type 555 timer to generate a square wave whose mark-space ratio is variable. The value of mark-space ratio is decided by the setting of the potentiometer VR1, and the frequency is decided by the value of capacitor C along with the total resistance in the circuit connected to the IC.

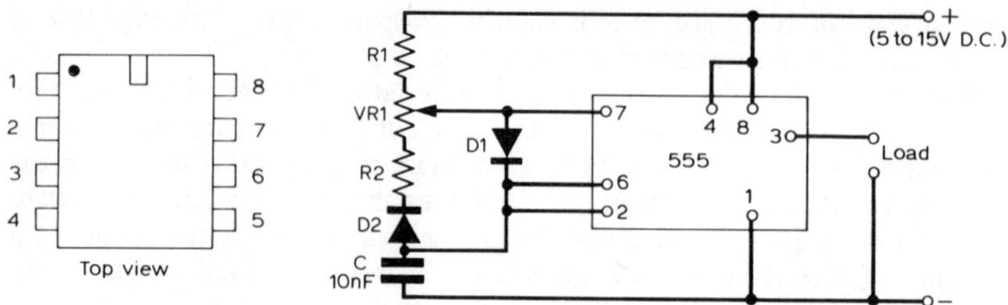

Fig. 3.3 Basic circuit for obtaining variable duty cycle, using a type 555 timer.

In the circuit shown, the values have been chosen to give a good range of mark-space ratio with a 1:1 ratio when VR1 is in the middle of its range. When the slider of VR1 is placed at the R1 end of the circuit, the mark-space ratio is very small, so that this end of the control is the low-speed, low power end. By convention, the wiring to the potentiometer is arranged so that this happens at the fully anticlockwise setting.

Operation of Squarewave Generator

Without going into details of the operation of the 555 timer circuit, the connections to pins 2 and 6 operate a switching circuit inside the timer which opens or closes a circuit between pins 7 and 1. When the voltage at pins 2 and 6 falls to just below one third of the applied voltage (which can be any value between 5V and 15V), then the circuit between pins 7 and 1 is opened, and the capacitor C starts to charge from the current which flows through R1, part of VR1 to the slider, then through D1.

This charging process causes the voltage across the capacitor to rise, until the voltage at pins 2 and 6 reaches two thirds of the applied voltage. At this level, the internal switch closes again, connecting pins 7 and 1 so that there is only a low resistance between them. The voltage at pin 7 is now earth voltage, with all the current passing through R1 and part of VR1 going to earth, and in addition the current which now flows out of C, discharging it, through D2, R2 and the other part of VR1. The two different current paths for charging and discharging are due to the fact that current will pass through a diode only in the direction shown by the arrowhead in the diode symbol.

Because the current paths are different, we can vary the amount of resistance in the two paths so as to make the charge time very different from the discharge time. With the slider of VR1 near the R1 end of the circuit, the charge time is short and the discharge time long; the position is reversed when the slider of VR1 is near the R2 end of the track.

At the output terminal, pin 3, the voltage is high while capacitor C is charging. During the discharge of C, the voltage at pin 3 is low. Due to the internal design of the IC, current can flow either from or into pin 3, and currents of up to 200mA may be used. This means that solenoids or small

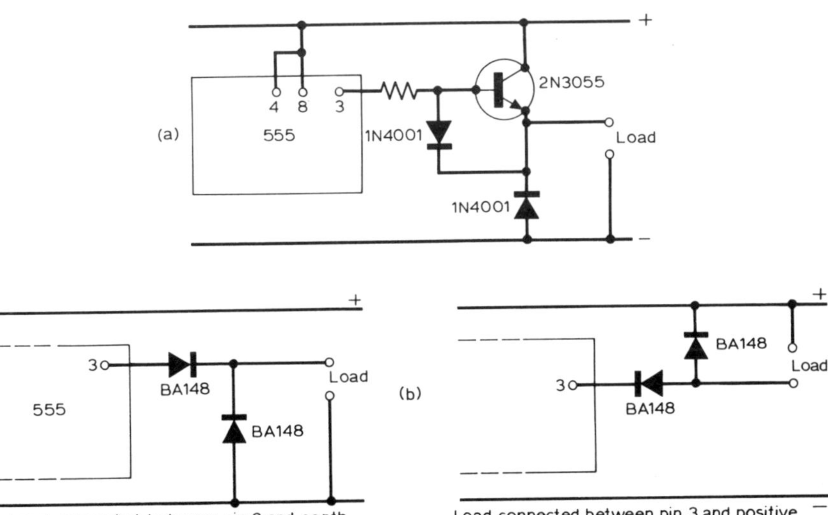

Fig. 3.4 (a) adding a power transistor to the basic circuit to provide more current (for a train controller using this type of circuit, see *Practical Wireless*, May, 1975), (b) using anti-latchup diodes with motor or solenoid loads.

motors of up to 200mA current consumption can be connected either between pin 3 and earth, or between pin 3 and the positive supply voltage line.

The circuit shown assumes that the load is connected between pin 3 and earth. If the load is connected between pin 3 and positive, then the maximum mark-space ratio is obtained when the slider of VR1 is near the R1 end of the track. In each case, if the output of the 555 timer is used to drive a relay, solenoid or motor directly, a diode must be connected across the connections of the solenoid, relay or motor to protect the IC from the destructively high voltages which can be generated when current is switched off.

In addition to these diodes connected in parallel with the load, a series connected diode must be used to prevent latch-up. A small negative voltage at the output pin, such as can be caused when current is switched off with a load connected (when the load is a motor solenoid or relay or other inductive load) can cause the 555 unit to stick with capacitor C fully charged and no switchover taking place. Such a "latch-up" can be cured only by switching off power for a short time and then switching on again; the problem is cured by using the series diodes, which prevent the voltage at the output pin from being driven negative by the load.

Variable-frequency Square Waves

The frequency of a squarewave can also be varied at the generator, and the changing of frequency can also be used to carry signals. We can use the 555

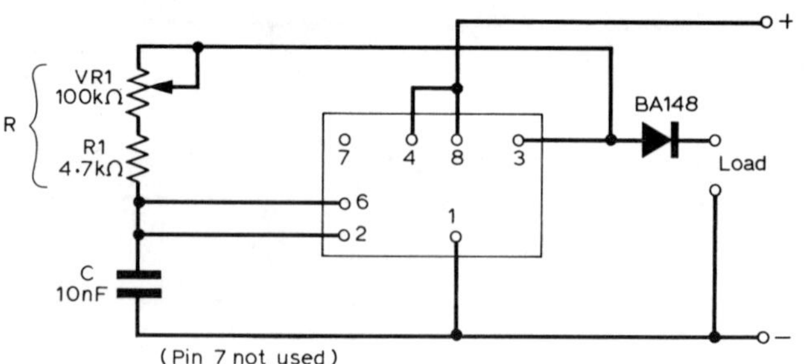

Fig. 3.5 Variable frequency generator, 50% duty cycle. If a fixed frequency is needed, VR1 and R1 can be replaced by a single resistor R (see text).

timer IC to generate squarewaves of 1:1 mark-space ratio and variable frequency by using the circuit of Fig. 3.5. Once again, pins 2 and 6 are connected together, and capacitor C is connected from pins 2 and 6 to earth. The capacitor is charged and discharged through resistor R, which may be fixed if a single output frequency is wanted, or variable for a range of frequencies.

When the output at pin 3 is high, current from pin 3 flows through R, charging up capacitor C so that the voltage across C rises. When the voltage across C reaches two thirds of the supply voltage the internal switch changes over, making the voltage at pin 3 low. The voltage at pin 3 is now less than the voltage across the capacitor, so that the capacitor now discharges back through R into pin 3. This continues until the voltage at pins 2 and 6 reaches one third of supply voltage, when the internal switch operates again, making the voltage at pin 3 high again. The process continues for as long as power is applied to the circuit.

The charging and discharging times are decided by the values of R and C. With values of resistance measured in kΩ and values of capacitance in nF, the times are given in us by the formula T = 0·7RC. For example, a 10nF capacitor and 47kΩ resistor will give a time of 0·7 x 10 x 47, which is about 330μS. Since the discharge time is equal, the time for the whole wave is 660μS. The frequency is 1/time, so that the frequency in the example is $1/660 \times 10^{-6}$ which is 1·5 kHz.

Applications of Timer

The wave generated by the timer should not be applied to any d.c. device directly, because the 1:1 mark-space ratio means that the device would simply operate on half power, and changing the frequency would have no effect. Small a.c. motors can be operated, however, and such small a.c. motors are now available, thanks to the fashion for electronic controlled watches, which now seems likely to disappear in favour of electronic digital watches. The motors used in tape recorders and disc players are usually of too high power for a 555 to operate directly.

Fig. 3.6 Coupling the a.c. signal in and out of rails on which d.c. is present.

The most suitable applications of this circuit are for switching solenoids, when the fixed frequency version can be used, and for some speed control applications, mentioned later. For solenoid control, the circuit of Fig. 2.12 can be used to convert the a.c. signals into d.c. We can carry the a.c. from the generator to the switching circuit of Fig. 2.12 along wires or rails which also carry d.c., provided that the signals are not shorted out by a low-resistance load.

In a model railway application, the presence of a train on line would normally bypass some of the a.c., but these high frequency signals are not so easily shorted out by a load of this kind, because the windings of the motor offer much greater resistance (more correctly, impedance) to a.c. than to d.c. Coupling the signals in and out by capacitors of about 10nF, with an a.c. frequency of about 5kHz, avoids such difficulties.

Frequency-to-voltage Conversion

We can make use of the variable frequency of signals from the circuit of Fig. 3.5 to carry information by converting frequency to voltage. A frequency-to-voltage converter takes in an a.c. signal of variable frequency and gives out a d.c. signal whose voltage is proportional to the frequency of the a.c. For example, we could have a circuit which had a d.c. output of 1V for a frequency of 200Hz in, 2V out for 1000Hz in, and 3V output for 1800Hz in.

This type of signalling can be an advantage when control voltages have to be carried along a path which has variable resistance, such as through a locomotive collector shoe or along the bearings of an RTP model aircraft pole. If we use different levels of voltage for controlling motor speeds or solenoid positions, then the effect of variable resistance is to change the speeds or positions in a way which is out of our control.

If, on the other hand, the signal quantity which we control is the frequency, and the receiver responds only to frequency signals, then the contact resistances will have little or no effect on the control voltages. This means that the signal must be converted to one which has constant amplitude but variable frequency; the receiver must then convert the different frequencies into different voltage levels to operate the control system. If this is being used to overcome contact difficulties, then the receiver must have its own power supply (battery), or some form of stabiliser which removes the fluctuations from the supply.

The best type of signal for a frequency control system is a squarewave with steep sides, because changes in the amplitude of the signal caused by contact variation do not cause any changes in the width of the square wave. The type of circuit detailed earlier in this chapter is ideal.

Squarewave Amplifier

At the receiver end of the control link, the squarewave may vary in amplitude because of the contact resistances, and an amplifier is needed to make the amplitude constant; the amplifier also serves in this case as a limiter. A suitable amplifier circuit is shown in Fig. 3.7. A single transistor is used, and is switched on by the positive peaks of the squarewave signal. Provided that the squarewave signal starts off with a reasonable amplitude, 5V or so, each wave into the transistor will switch the transistor on, even if the amplitude of the wave varies considerably.

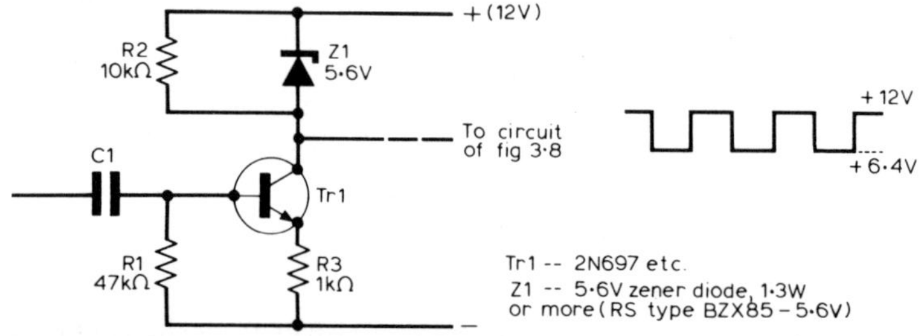

Fig. 3.7 Amplifier-limiter for a.c. signals. R3 is a protective resistor to prevent excessive current from passing through the diode and transistor.

At the output of the amplifier, each time the transistor is switched on the voltage level falls from supply voltage to 5·6V lower, a level which is held by the zener diode. Greater amounts of current cause no change in this voltage (this is the action of the zener diode), so that the output of the amplifier is a set of squarewaves whose peak-to-peak voltage is 5·6V.

Conversion Circuit

The conversion of frequency signals to voltage signals can be carried out by Fig. 3.8 circuit. The squarewave input passes through the capacitor C, and the diode D1 conducts when the voltage of the wave is below earth voltage. Since the mid-point of the wave would normally be at earth voltage after passing through the capacitor, this causes some charge to accumulate in the capacitor, so that we would expect the plate connected to the diode to reach 5·6V above earth when the wave reached its positive peak.

At positive voltages, however, the second diode, D2, can conduct, so that some of the charge on C1 is shared with C2. Charge then flows from C2 into the base of the transistor, so that C2 has to be topped up all the time from C1,

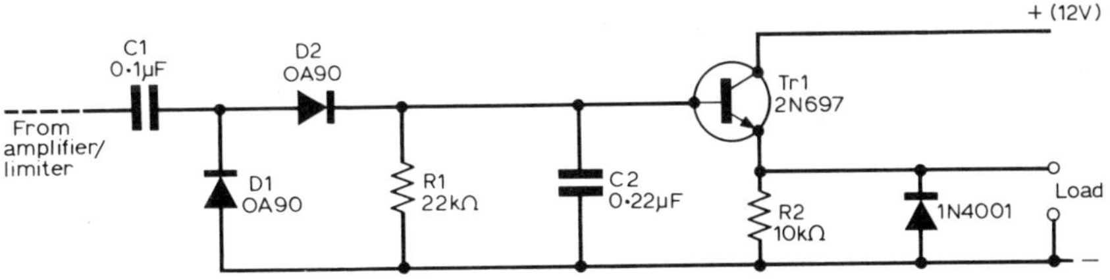

Fig. 3.8 Frequency-to-voltage conversion circuit.

which is smaller than C2. This topping up occurs more often when the square-wave is at a high frequency than when it is at a low frequency, so that the current flowing depends on the frequency. In the circuit shown, using values of 0·1μF for C1, 0·22μF for C2 and 22kΩ for R1, the voltage across R1 with no transistor connected is about 1·1V at 1kHz, down to 0·2V at 100Hz.

The connection to the transistor will depend on the use which is being made of the circuit. If a large range of frequency is wanted to produce a small voltage change, the output can be taken from the emitter of the transistor whose base is connected to C2. For greater sensitivity, the output can be taken from the collector, where a greater range of voltage is available. A useful circuit is shown in Fig. 3.9. The resistor in the emitter lead prevents the circuit from being too sensitive, and the variable VR1 adjusts the value of the output voltage at the lowest frequency used.

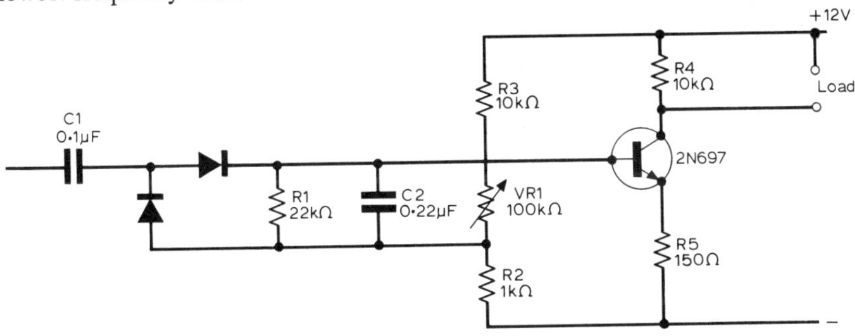

Fig. 3.9 A more sensitive frequency-to-voltage conversion.

Time Delayed Switching

In many applications, a delayed switching is useful; a signal which indicates clear road shortly after points have been set, or a motor which stops after exactly 50 seconds of running. Such time delays can be carried out electronically, and the range of times which is now available is very large. In addition, it is now possible to make use of lightweight batteries to construct units lighter than the mechanical or pneumatic timers which have been used for such purposes as aircraft dethermalisation. At the other end of the time scale, the timer

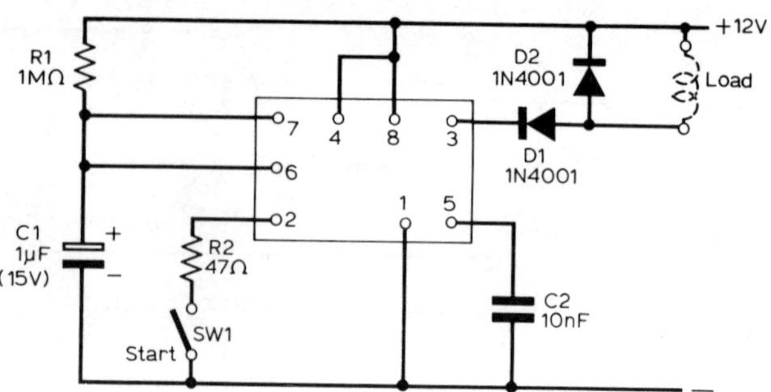

Fig. 3.10 Using the type 555 timer circuit as a time delay. Basic circuit, with values of R1 and C1 selected to give a delay of about 1·1 seconds. The output drives a solenoid or similar load directly.

can be used to ensure that one operation is always carried out a set time after the preceding one, with a lockout so that the second operation can never be carried out first.

For timing delays, the 555 timer unit which we have already used is ideal; this is the purpose for which it was designed. A simple delay circuit is shown in Fig. 3.10. Here a type 555 timer is being used to drive a d.c. load such as a solenoid or motor, represented by a coil, and protected by diodes.

The timing is started by switch Sw1 which links pin 2 to earth. This switch must not remain on, because timing will start at the instant the switch contacts meet but will not stop if the switch remains on at the end of the timed period. Sw1 should therefore be either a pushbutton switch, or a switch fitted to a capacitor as shown in Fig. 3.11a, which delivers one pulse and then leaves pin 2 free to take up its own voltage level.

A slightly modified version of this circuit can be used for touch-start operation, in which the timing is started simply by touching a metal strip of contact with one finger. The 100kΩ resistor, R1, in Fig. 3.11b, controls the sensitivity of the starting circuit, so that if the starting proves to be too sensitive, triggering on approach or when brushed by clothing, the value of R1 can be reduced to 27kΩ or less.

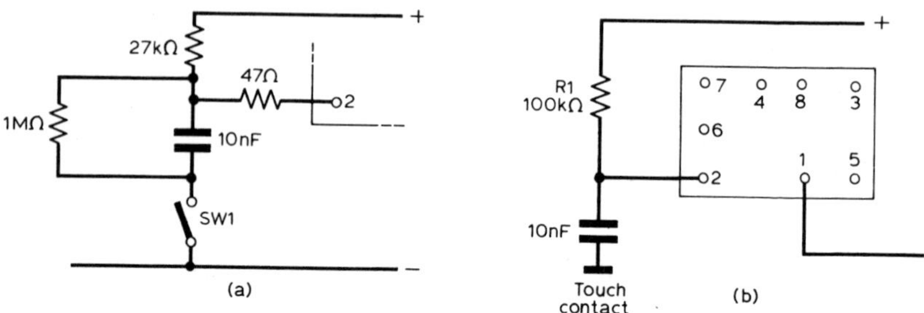

Fig. 3.11 Alternative starting circuits. (a) switch circuit which will deliver a single pulse, (b) touch switch starting.

Long Delays

Delays of up to 18 minutes are obtainable using the 555 timer, but some care has to be taken with the selection of suitable components for such comparatively long times. The delay time is determined by the value of R1 and C1 in Fig. 3.10, so that delay time in seconds is given by 1·1R1C1, with R1 in units of MΩ and C1 in μF. For example, with a resistor of 1MΩ and a capacitor of 1μF, the time would be 1·1 seconds. For a 10MΩ resistor and a 5μF capacitor, the delay would be 55 seconds.

There is a limit to the size of resistance and capacitance values which can be used, however. The value of R1 should be less than 1·3 x V (in MΩ), where V is the voltage of the positive supply to the timer unit. For example, if we use a 9V supply, the value of R1 must not be greater than 1·3 x 9 = 11·7MΩ. In practice, we would not use more than 10MΩ, which is the nearest available value.

Capacitor values of 10 000 μF or more are available, but capacitors of very large value usually have low values of permitted voltage, and in addition allow comparatively large amounts of current to leak through. This restricts the size of capacitor values which we can use, unless expensive low-leakage tantalum capacitors are used.

A value of about 100μF with a working voltage of 15V is quite reasonable, however, and can be obtained in conventional construction; this combined with a 10MΩ resistor gives a time of 1100 seconds, just over 18 minutes, which is long enough for most purposes. New designs of timer units are capable of much longer delays, but use integrated circuits which have to be handled with much greater care during circuit construction. For that reason, they will not be discussed here.

CHAPTER FOUR

ELECTRONIC MEASUREMENTS

MEASUREMENT IS AN ESSENTIAL PART of modelling, and one which can be aided considerably by the use of electronic circuits. Before we look at circuits which can be used for purely modelling purposes, however, a measuring instrument for checking electronic circuits themselves is extremely useful, and can be used also for electrical checks in modelling applications. Voltage readings are frequently needed in checking electronic circuits, whether for the circuits in this book or for the more elaborate circuits used for radio control equipment; and also for checking batteries, rail supplies, glow-plug accumulators and other uses.

There is an important difference between the readings made on electronic circuits and those made on simple electrical circuits, however. The measurement of a battery voltage can be made by any meter of reasonable accuracy, because the battery is capable of supplying the current which the meter must take in order to operate. The measurement of voltages in electronic circuits needs meters which take very little current from the circuit which is being

Fig. 4.1 Specimen voltmeter and ammeter.

tested, so that the use of the meter disturbs the action of the circuit as little as possible.

Such meters are rather expensive to buy, but a very ordinary current meter of 1mA range can be converted to read 10V full scale, and take only a very small amount of current from the circuit which is being tested, less than 10μA. Similarly, an older type of voltmeter or multimeter can be converted so that it takes much less current and is therefore much more suitable for electronic uses.

Converting a 1mA Meter

The circuit used for the conversion of the 1mA meter is shown in Fig. 4.2. A single type 741 integrated circuit amplifier is used to drive current from the battery through the meter, M, to the junction of the two 9V batteries. The resistors R1 and R2 are arranged so that the voltage gain of the amplifier is nearly ten times, and the input resistors R3 and R4 scale down the input voltage (up to 10V) so that the input to the amplifier is not overloaded.

Two adjustable resistors, VR1 and VR2 have to be included, but seldom need any alteration, so that they can be small preset "skeleton" types. VR1 corrects for "offset", so that with the input terminals shorted together, the value of VR1 can be adjusted so that the needle of the meter is exactly on the zero mark. VR2 corrects the range of the meter, and has to be adjusted with a known calibrating voltage. This can best be provided by another 9V battery and a voltmeter of known accuracy, such as the Avo Mk.8 or similar instrument. If a good quality meter cannot be borrowed for calibration, use a fresh 9V cell connected to the input terminals and adjust VR2 so that the meter reads 9V.

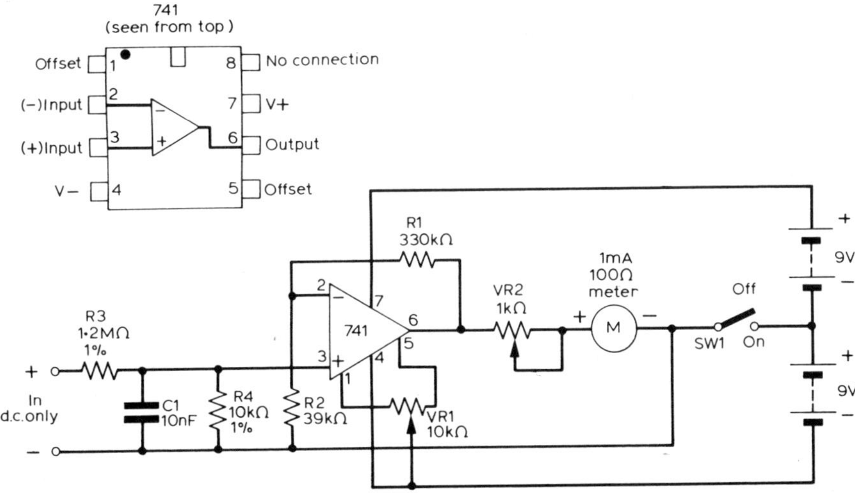

Fig. 4.2 Adapting a 1mA meter so that the input resistance is high.

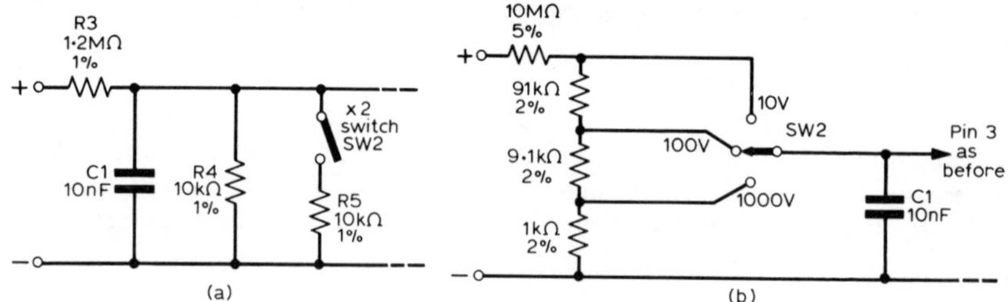

Fig. 4.3 Adding ranges. (a) x2 switch, extending the meter range to 20V, (b) multi-range meter with ranges of 0-10V, 0-100V and 0-1000V d.c.

The meter calibrated in this way is now ready for use on most of the electronics circuits described in this book. For those circuits which use voltages of more than 10V (some supply voltages), a x 2 range can be fitted as shown in Fig. 4.3a. If a wide range of voltages have to be measured, a set of ranges from 10V to 1 000 V d.c. can be obtained using the input circuit of Fig. 4.3b. These ranges are approximate because of the values of resistors which can be easily obtained, but are nevertheless accurate enough for our purposes. A.c. ranges are seldom needed for modelling purposes.

The construction of the amplified meter is fairly straightforward, since all the electronic components can be mounted on a board, with a metal case to mount the meter and switches, along with the batteries. Battery holders can be obtained from electronics stockists. The on/off switch should preferably be of the type which is spring loaded to the **off** position, so that the switch has to be pressed on for a reading. In this way, the instrument cannot be left accidentally switched on, and the battery life should be at least a year.

Modifying a Voltmeter

If an old voltmeter with a suitable range is available, it can be modified in the same way with no alterations inside the meter, but with the restriction that the voltage range on the meter must be within the 9V limit set by the batteries which are used in the amplifier. The meter will be suitable as long as its resistance is greater than 200Ω, since the 741 must not be used to supply loads of less than 200Ω.

This information can usually be read from the meter scale in some form or other. One form is the figure of "ohms-per-volt". This means the resistance of the meter divided by the scale used. For example, a 100 ohm-per-volt meter used on its 6V range would have a resistance of 6 x 100Ω, 600Ω, and suitable for our purposes.

The other way of stating resistance is to quote the meter current at full scale deflection, usually abbreviated to f.s.d. current. If the full-scale voltage is divided by this current, the result is the resistance. If the units of current are mA, the resistance is in Ω; if the current is in amps, the resistance is in ohms.

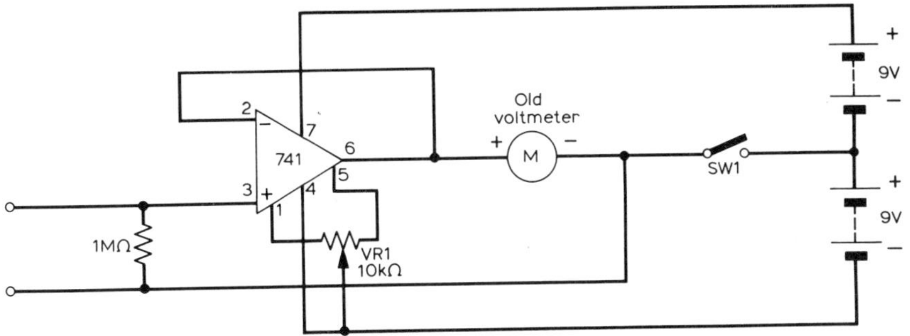

Fig. 4.4 Using an old d.c. voltmeter with an amplifier.

For example, a meter of 15mA f.s.d. on its 2·5V range has a resistance of 2·5/15kΩ, which is 0·166kΩ, or 166Ω, too low for our purposes.

If the voltmeter is suitable, we can use a type 741 amplifier in the circuit of Fig. 4.4. As it stands, the gain (output voltage / input voltage) of this circuit is unity, so that 6V in will give 6V out, with the important difference that the input current will be only a microamp or so, but the output current can be 30mA or so, enough to drive the older types of insensitive meters. Two modifications can be made to change the range of amplifier. One possibility is to increase the amplification by the circuit of Fig. 4.5.

The amplification is set by the ratio of VR1 + R2 to the total of VR1, R2 and R3, so that alteration of the value of VR1 will change the amplification. In this way, a 6V basic meter movement can be made to read full scale on 3V, or 5V or whatever suitable quantity is needed. Note that this does not make it possible to use a meter intended for a full-scale deflection of more than 9V, since the current to operate the meter comes from the batteries of the amplifier.

The other possibility is to extend the range to higher voltages by using a voltage divider in the input circuit as before. Because of the very small currents

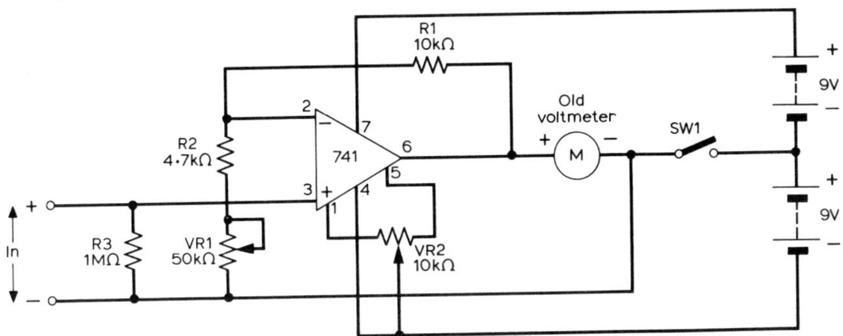

Fig. 4.5 Extending the range of an old voltmeter. The values have been chosen so that the amplification of voltage is up to 3 times.

taken by the amplifier, large value resistors can be used, so that the current taken is still very small. Note that these alterations to the range of the meter may need some mental gymnastics when a reading is taken. For example, if we have altered a 6V meter so that it reads 10V full scale, each reading on the scale must be multiplied by 10/6 (which is 1·67) to get the true reading. In addition, calibration is again needed when the range is set by altering VR1 in the circuit of Fig. 4·5

Using Transducers

Transducers are devices which convert various measurable quantities into electrical signals of some sort, or the other way round. A microphone, for example, is a transducer of sound signals into electrical signals; a photocell is a transducer of light signals into electrical signals. By making use of transducers, we can make electrical measurements of quantities which would otherwise be difficult to measure.

One good example of the use of transducers is the measurement of comparative wind speed for aero and yacht modellers. To measure wind speeds accurately with any instrument needs access to a wind tunnel for calibration, but a fairly simple wind speed meter can be made which is self-consistent, can be calibrated in the numbers of the Beaufort Scale, and which will thereafter give useful comparative readings.

The transducer here is a miniature electric motor using a permanent field magnet. Such motors are widely used in slot-racing cars and in RTP aircraft models. These motors will also act as generators, and the generated output, which will be a rough d.c., is proportional to the speed of the motor shaft over a good range of speeds. The output is small, but it can be amplified electronically. The shaft can be driven by the wind by attaching an anemometer head to the shaft, consisting of a piece of dowel about 15cm long with cups made from halved table-tennis balls at each end. A de-luxe version can be made using 4 cups. The advantage of this type of system over a propeller is that it is not sensitive to wind direction, so that the assembly does not have to be placed on a hinged mount. In addition, the response to low wind speeds is better. The assembly should be carefully balanced and then glued, using a quick-set epoxy

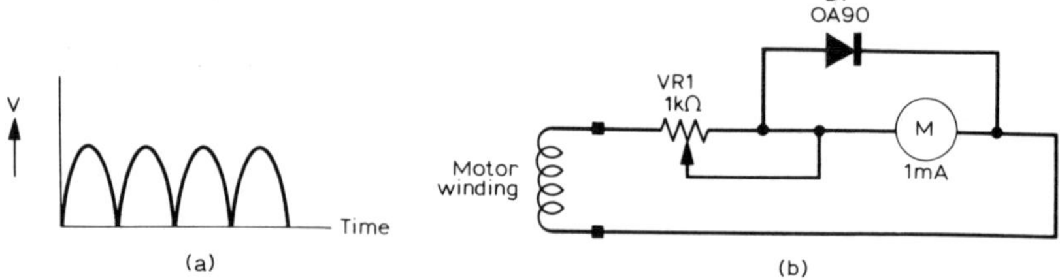

Fig. 4.6 Using a small motor as a wind-powered generator. (a) output voltage, (b) simple indicator circuit.

resin, to the shaft of the miniature motor, so completing the transducer part of the assembly.

This simple transducer will give a waveform of the type shown in Fig. 4.6a, and a voltage output ranging from 0·04V in calm conditions to about 1V in a gale. Such a range of outputs can drive a 1mA meter directly, with no amplification needed, so that the simple circuit of Fig. 4.6b can be used. The variable resistor VR1 is set so that the meter reads full scale in conditions which are just too windy for model flying. The diode, D1, prevents damage to the meter caused by excessive currents generated in gusts.

Since 1mA meters are now rather expensive items, some alternative may be needed. If an old meter movement is available, whether current or voltmeter, it can be pressed into service by using the circuits of Fig. 4.2 to 4.5 to adapt the range of the meter. In each case, the variable resistor is used to set the range of the complete instrument so that the meter reads full scale in conditions which are just too windy for flying.

Using LEDs

Since meters are comparatively fragile and heavy, we can make a cheaper and simpler instrument if a go-nogo type of indication is satisfactory. We can achieve this type of indication by using the generator to provide the input to an amplifier which operates miniature light sources called light-emitting diodes (LEDs). These LEDs are solid, with no fragile filaments, and have a long working life, using low voltages and currents. Using one green and one red LED, the instrument can be set up so that safe wind conditions are indicated by a green light; unsafe by a red light, and doubtful conditions by both lights being illuminated.

The complete circuit of the go-nogo indicator is shown in Fig. 4.7. The signal from the transducer is taken to the input of the transistor, and the earthy end of the input is taken to VR1, so that a steady voltage can also be applied to the input of the transistor. The voltage applied to the emitter of the transistor by resistors R1 and R2 makes sure that a voltage of about 1V from the transducer is needed to turn on the transistor. This would normally be available only in a gale, so the variable VR1 is used to set the wind level at which the transistor starts to pass current. When the signal from the transducer is too low to turn the transistor current on, the voltage at the output of the transistor (the collector) is high, and current will flow from the positive line through R3, R4, D1 into the green indicator LED. Note the symbol for the LED, which is similar to the symbol for a diode, because the construction is very similar.

As the wind strength increases, spinning the shaft of the transducer faster, the output voltage increases, and the transistor will be switched on by the peaks of the voltage pulses. The capacitor C1 smooths out these peaks to some extent, and the red light does not come on until the voltage at the output is several volts down (on average) from the supply voltage. When this happens,

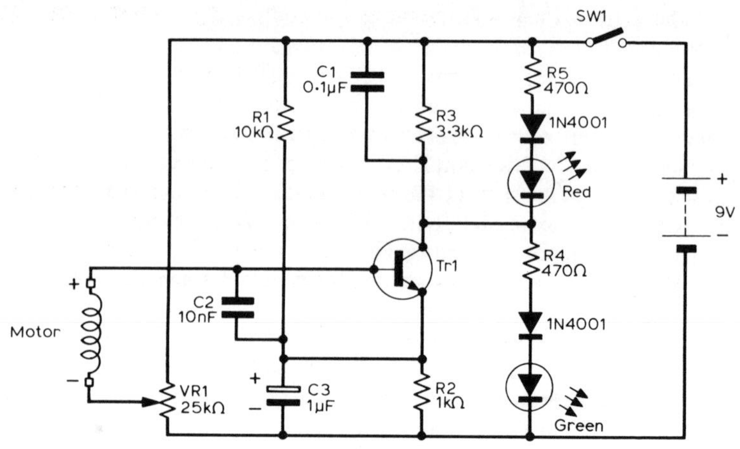

Fig. 4.7 Wind indicator using a Go-NoGo indication.

current can flow through R5 and D2 to the collector of the transistor, so that the red LED is illumintated. For a range of voltages at the output, both indicators will be lit, but when the voltage at the collector falls below about 3V, the green LED is extinguished.

Because of the small size of the components, and the fact that only one battery is needed, this indicator can be built in a much more compact form than using a meter. Calibration consists of adjusting VR1 on a day when the wind is just too great for flying, so that the green indicator is just extinguished. Thereafter, the indications will be consistent, and should be considerably more reliable than the time-honoured methods of pointing wet fingers or throwing up dry grass.

The electric motor used in this way (the term for such a transducer is tachogenerator,) can be applied to the speed measurement of revolving shafts, with the output speed read from a meter, but the power absorbed by the tachogenerator is a noticeable load for a small power unit, and other methods can be used. For marine engines, steam or internal combustion, however, this can often be a very convenient method of speed measurement, though calibration against a rev. counter (tachometer) is needed.

Exhaust Noise Meter

Wherever model engines are run, noise regulations stipulate that suitable silencers must be used, and sound level measurements can be applied to enforce the noise laws. Though the construction of a sound level meter of the type which would satisfy the legal requirements is beyond the scope of this book (and beyond the understanding of a lot of anti-noise campaigners), a simple sound level indicator can be constructed which enables the effect of different silencing systems to be compared.

The metering system is composed of a microphone, which is the transducer, an amplifier, and a rectifier which converts the a.c. output of the amplifier into d.c. which can be read on a meter. The microphone used can be a cheap dynamic microphone insert which can be obtained at low cost at the time of writing. This is used to pick up the noise level at a set distance from the source by building the microphone amplifier and display unit into a box fitted with a 50cm probe. In use the end of the probe is placed against the motor housing, so that the microphone is 50cm from the noise source.

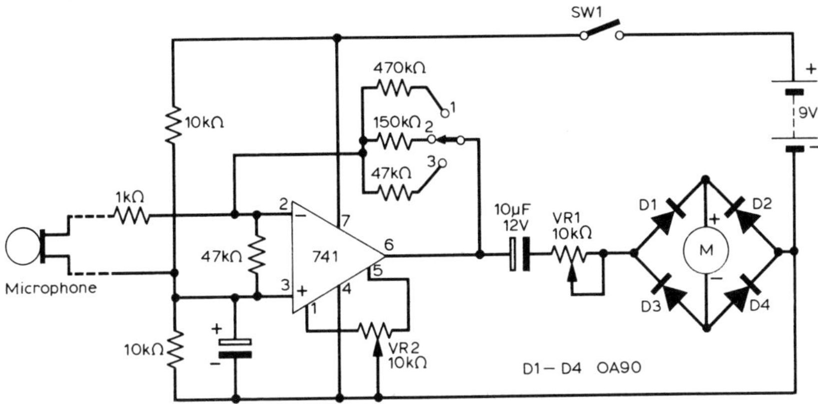

Fig. 4.8 Sound level indicator. Before calibration (see text) VR2 should be set by measuring the voltage at pin 6 of the IC and adjusting VR2 so that the reading is 4·5V.

The output of the microphone, which is an a.c. signal, is amplified by an integrated circuit amplifier whose gain can be switched to several ranges, and the a.c. output of the amplifier is rectified and used to drive a meter. Because the characteristics of microphones differ so greatly, and are affected by the mounting arrangements used, no precise values can be given for the ranges, but the suggested values in Fig. 4.8 should cover the sound outputs normally found from model motors of various types at 50cm spacing.

Since meters are rather expensive, the indicator may be adapted for use with a separate meter by taking the meter terminals out to sockets on the case. In this way a meter which normally earns its keep attached to a wind-speed indicator, or a voltmeter, can be used for the sound level comparator as well.

The most useful calibration for this type of insttument consists of removing the silencer from an engine which is known to be particularly noisy, starting up in a place suitably remote, and using the meter with the engine going at full blast. The variable range adjuster, VR1 is then set so that the meter reads full scale on its least sensitive range. The other ranges are now calibrated automatically, because the values of the resistors used in the circuit will keep the readings of the other scales consistent.

This type of sound level meter is more sensitive to changes in sound level than either the human ear or the expensive professional sound level meter. This is because the scale of our meter is (approximately) linear, meaning that when the sound level changes from maximum to half of maximum the meter reading also changes from maximum to half of maximum. The ear has a much more complicated response, approximately logarithmic, in which a change of ten times in the sound level produces a change of around two times in the loudness. To complicate matters further, the ear is much more sensitive to some frequencies (those most noticeable in human speech, logically enough) than to others, so that professional sound level meters have to be "weighted", meaning that the reading on the meter depends on the frequency of the sound as well as on its intensity.

Bridge Measurements

One very common type of measurement used in electrical and electronic work is the bridge measurement. Bridge measurements compare quantities in a way which gives a voltage or current reading of zero when the quantities match, so that elaborate or expensive meters are not needed. The disadvantage of a bridge measurement is that the reading cannot be simply read from a dial unless the equipment is calibrated beforehand.

The simplest form of bridge is the Wheatstone bridge used to measure electrical resistance. The basic Wheatstone bridge circuit, familiar to generations of students of "A" level Physics, is shown in Fig. 4.9. When the ratio of values R1/R2 exactly equals the ratio of values R3/R4, then the current through the meter is zero, and the bridge is said to be balanced. If we know the value of R1 and the value of the ratio R3/R4 (without necessarily knowing the values of R3 or R4), we can find the value of R2.

No readings are needed from the meter, which is used only to show when current is zero, and the accuracy of the measurement is as good as the accuracy of the value of R1 and the ratio R3/R4. On a practical instrument, a switch is used to select different value of R1, and a potentiometer is used to set value of the ratio R3/R4 from 0·1 to 10, so that values of R1 can be read quickly when the bridge is balanced.

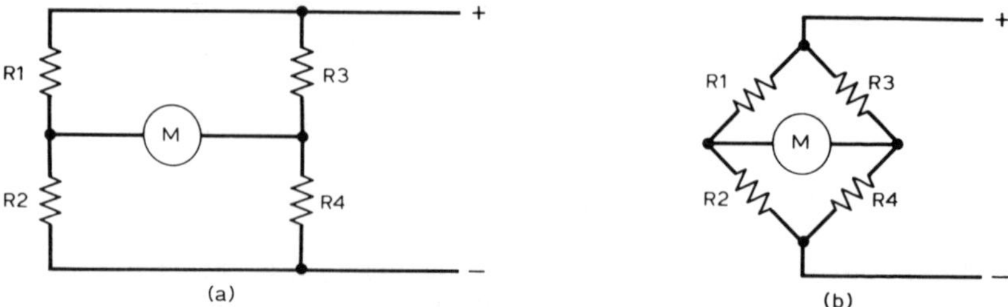

Fig. 4.9 Wheatstone bridge network. (a) circuit diagram, (b) alternative way of drawing the circuit.

The simple bridge is satisfactory only if suitable meters or indicators are available, and a much greater range of use is possible if an amplifier can be used. The integrated amplifiers which we have previously used are ideal for this purpose, as they operate by comparing the voltages applied to two inputs. When the voltages applied to the two inputs (coded + and −) are exactly equal, the output of the amplifier is exactly zero volts when exactly equal positive and negative supply voltages are used. A small difference between the two input voltages causes the output voltage to change considerably. Because of the high gain of these amplifiers, we do not need to use a sensitive meter for deciding when a bridge is balanced. Instead we can use two LEDs, with a red LED indicating too high a value of unknown resistance for the settings used, a green LED indicating too low, and balance when both LED's are illuminated.

Resistance measurements can often be of direct use in modelling work. The contact resistance between collector shoes, pantographs or locomotive wheels and their respective rails or wires is one example of resistance which can be measured; the change of resistance when points are switched is another. Looking at a different activity, the resistance of wires used for RTP model flying is of importance if the correct current is to be delivered to the motor. Slot car racing also has applications for the measurement of contact resistance.

Bridge Circuit

A measuring bridge circuit is shown in Fig. 4.10. The bridge itself consists of the variable potentiometer VR1, a 10kΩ *linear* potentiometer, and the fixed value resistors R1, R2, R3 which can be switched into circuit. The unknown resistance is wired between the terminals X and Y. The voltage between R1,

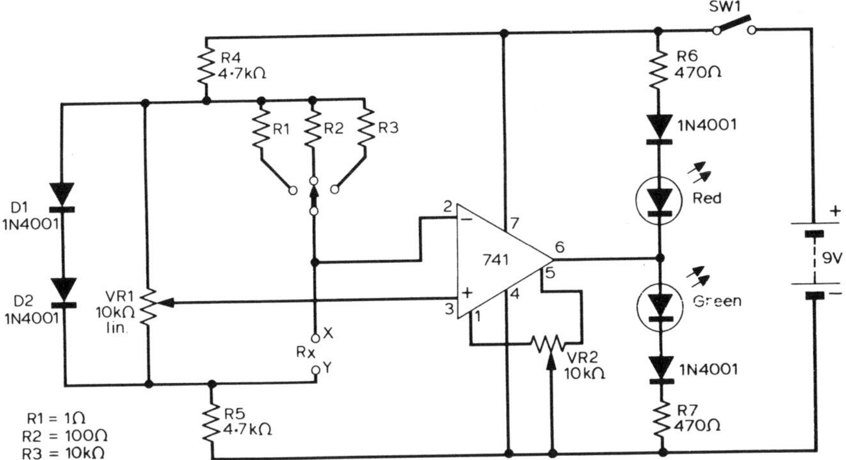

Fig. 4.10 Measuring bridge circuit using LED indicators. The voltage across the bridge is small, and is set by D1 and D2.

2, or 3 and the unknown is compared with the voltage at the slider of the potentiometer VR1 by taking each voltage to an input of the amplifier, a type 741. When the input voltages are made exactly equal by adjusting VR1, the output of the amplifier will be exactly between the voltages of the + and − lines, and each LED will glow.

Because of inevitable differences in the amplifiers, and also because the voltage across a red LED needs to be rather less than that across a green LED, the output voltage has to be offset, and this is done by adjusting VR2 so that with a resistance connected across XY which is exactly equal to R2, and with the switch selecting R2, and VR1 exactly half way round its track, both LEDs are equally illuminated.

With the null adjustment made, the bridge can now be used for measurements. The resistance to be measured is connected across XY, the range switch set to an appropriate range (lowest for contact resistances) and the potentiometer adjusted until both LEDs are lit. If this happens only at the extreme of the potentiometer range or not at all, then another position of the range switch must be used. At balance, the value of the unknown resistor is obtained by multiplying the reading on the potentiometer scale by the figure on the range switch.

For example, if the switch were set to 100Ω and the potentiometer read 0·3 at balance, the resistance connected across XY is 0·3 x 100 = 30Ω. The three ranges shown make it possible to measure values of resistance from 0·1Ω to 100kΩ; another range can be added (using a 1Mohm resistor) which can cover the values of 100kΩ to 10MΩ if needed.

Calibrating the Potentiometer

When the circuit is built, the potentiometer has to be calibrated. Scales can be bought which have the required numbers printed on them, with 0·1 at one extreme and 10 at the other, and 1 in the centre, but a more precise calibration can be obtained if a blank circular scale is made up and the calibration marks filled in. A pointer knob is fitted to the spindle of the potentiometer VR1, and the dial is attached to the case temporarily, but in such a way that it can be taken off and replaced permanently in exactly the same position.

With the dial in place, set the range switch to 100Ω and connect another 100Ω resistor across XY. With the unit switched on and the null adjustment correctly set as detailed earlier, adjust VR1 for balance, which should be at the centre of the range of the potentiometer. Mark the position of the pointer on the dial, which will be the "1" position. Now repeat using a 1000Ω resistor, balance, and mark this position on the dial "10". Using a 10Ω resistor, balance and mark the dial "0·1". The dial can now be removed and the remaining steps between these marks filled in.

Because the potentiometer is linear, the space between the divisions can be ruled off into equal subdivisions, using a large protractor. Measure the angle between the "0·1" and the "1" marks. Divide this by 9, and mark off at this value of angle all the divisions between 0·1 and 1·0. For example, if the angle between 0·1 and 1·0 is 81°, then there will be 9° to each division, so that the mark for 0·2 is 9° from 0·1, 0·3 is 9° beyond 0·2 and so on. Repeat this for the steps from 1·0 to 10, and the dial is ready for use. A coat of transparent protective lacquer is a good precaution before the dial is replaced.

Temperature Measurement

The basic bridge circuit can also be used for temperature measurements over a wide range, including steam temperatures. For temperature measurement, the resistance between X and Y is replaced by a thermistor, which is a piece of semiconducting material which is very sensitive to temperature changes. Unlike metals, the material used for thermistors has a negative temperature coefficient (NTC) which means that the resistance decreases as the temperature is raised.

The peculiarity of the material used for thermistors is that the change of resistance is very considerable, perhaps from thousands of ohms to less than a hundred ohms for the change of temperature from room temperature (about 20°C) to that of water boiling under normal pressure (100°C). Since thermistors can be made in very small sizes, their use for temperature measurement means that temperatures can be measured in places where conventional thermometers could not be placed, such as on the fins of aero engines, and along steam pipes. Their sensitivity means that small changes in temperature can be measured, so that the drop in steam temperature as steam passes from boiler to cylinder can be measured.

In addition, since the thermistor is an electrical transducer the indicating part of the thermistor thermometer (the bridge circuit) can be some distance away from the thermistor itself, and one bridge can even be used to take readings from a number of thermistors so that several temperatures can be taken in succession by switching different thermistors into circuit. When this is done, each thermistor has to be calibrated, so that the bridge should be used to find the resistance, and the reading of temperature taken from a calibration chart for the thermistor in use.

Thermistor Bridge

A simple thermistor bridge, such as can be obtained by connecting a small glass-bead thermistor (such as the RS Components THB12) between points X and Y on the measuring bridge, using the 1kΩ range, must be calibrated. One simple method is to tie the thermistor to the bulb of a wide range glass thermometer (−10°C to +300°C thermometers are available), and take readings of resistance and temperature over a wide range, so that a calibration graph can be drawn. Avoid getting the leads of the thermistor under water; a sand-bath is preferable to a water bath for this calibration and has to be used for temp-

Fig. 4.11 Calibrating a thermistor.

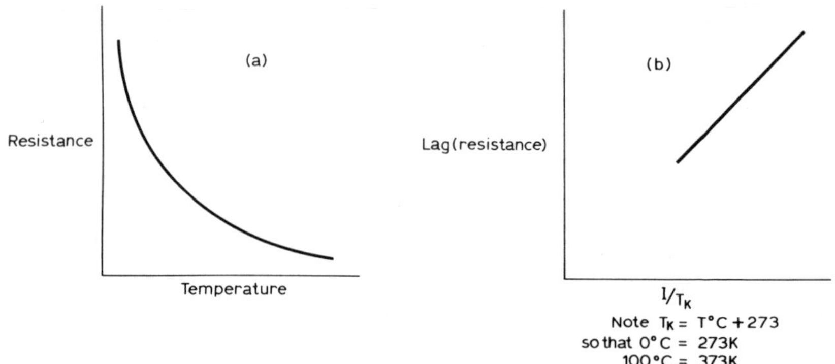

Fig. 4.12 Calibration graphs.

eratures above 100°C anyhow. A more useful graph to plot is of logR against $\frac{1}{\text{temperature(X)}}$, which gives an approximately straight line, and makes it easier to extend the range beyond the calibrated points.

Some care has to be taken, particularly with miniature thermistors smaller than the type noted above, that the current passed through the thermistor is small. Excessive current will cause the thermistor to heat up, just as any other resistor would, so that the temperature measured is not the true temperature. The bridge circuit passes fairly low currents, depending on the value of range selected; but the 10kΩ range is preferable for the higher resistance thermistors.

CHAPTER FIVE

COUNTING CIRCUITS

COUNTING IS AN ACTIVITY which can be particularly easily carried out using electronic circuits, which is why calculator and computer circuits are based on counting circuits. We can count the total (e.g. the total number of laps in a race) or a number occurring in a set time (such as the revolutions per minute of a propeller) or the amount of time between one event (start) and another (stop) by using different arrangements of circuits.

The units which are used for electronic counting are pulses, sudden changes of voltage (Fig. 5.1). To avoid having to refer to exact voltage values, we usually call one of the voltage levels, the lower value, zero, and the other one, 1. In this way we need only use the numbers 0 and 1 when we refer to the state of a pulse, and we can arrange the circuits so that the precise voltages are unimportant. For example, we can arrange the circuits so that the 0 level can be anywhere between zero volts and 0·4V and the 1 voltage can be anywhere between 2·4V and 5·25V. These example voltages are, in fact, used in many counting and computing circuits. We must, of course, have some method of turning whatever quantity we want to measure into these pulses.

Digital and Analogue Counting

Counting pulses in this way is called digital counting, and the most startling developments in electronics in the last few years have centred on digital methods. It is not, however, the only method of electronic counting. It is also

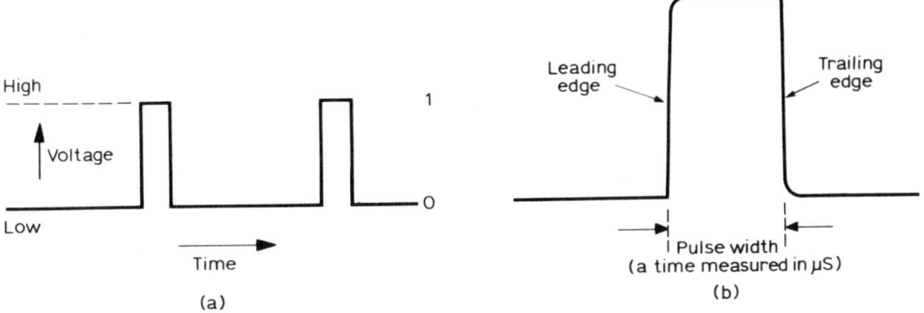

Fig. 5.1 Pulses. (a) pulse shape, (b) detail of pulse.

possible to use each item which is to be counted to increase the level of voltage which can then be indicated on a meter, so that instead of dealing with two voltage levels we are using as many different values as we have figures to count.

This system is called analogue counting, and can be used to make very simple counter circuits. Our wind-speed indicator, for example, used analogue counting, for the wind speed was converted to a series of pulses by the tacho-generator, and these pulses were "counted" by the metering circuit to give a meter reading proportional to the number of pulses per second and the amplitude of the pulses.

The differences between the systems are important. For small scale electronic circuits, such as we mainly use in this book, analogue methods are often simpler and better suited to anyone more interested in results rather than methods. Digital methods are sometimes more useful, certainly more accurate, often much more complex and expensive. We shall use whatever methods are simplest and cheapest in each example, rather than devise elaborate circuits which would be the equivalent of using a sledgehammer to crack a nut.

Lap Counting

One obvious application of counting is the counting of laps for slot-car racing. This is easy enough to carry out if only two cars are being raced, but becomes rather a struggle for one lap judge if a large number of tracks are in use. An automatic lap counter for each track makes the job much easier, and avoids disputes. The simplest and most satisfactory way of recording the lap count is to use an electromagnetic counter which clicks forward by one digit for each pulse into the coil of the counter. These counters are comparatively cheap, and their main drawback for most electronics applications, high power consumption and low operating speed, is not of importance when mains supplies (or accumulator supplies) are available and for low-speed work like lap counting.

Most electromechanical counters will operate up to 10 counts per second, some faster, so they are not exactly working to their limits on slot racing. The main problem, therefore, is to generate the pulses for the counter. Mechanical methods, such as pressure switches operated by the weight of the car, are the most straightforward electrically, but are also the most difficult to set up and the most difficult to maintain in good working order. The best methods from the points of view of simplicity and minimum maintenance and modification to the track and cars, are the light beam and the magnetic pickup.

Light-beam Method

The light-beam method uses a small light bulb or LED to project a beam of light into a photocell (an electronic light detector) which is hooded so that it is not affected by other lights which may be present. When the beam of light is interrupted by the object which is being counted, such as a car in this example, the resistance across the photocell increases greatly, and this can be used in a

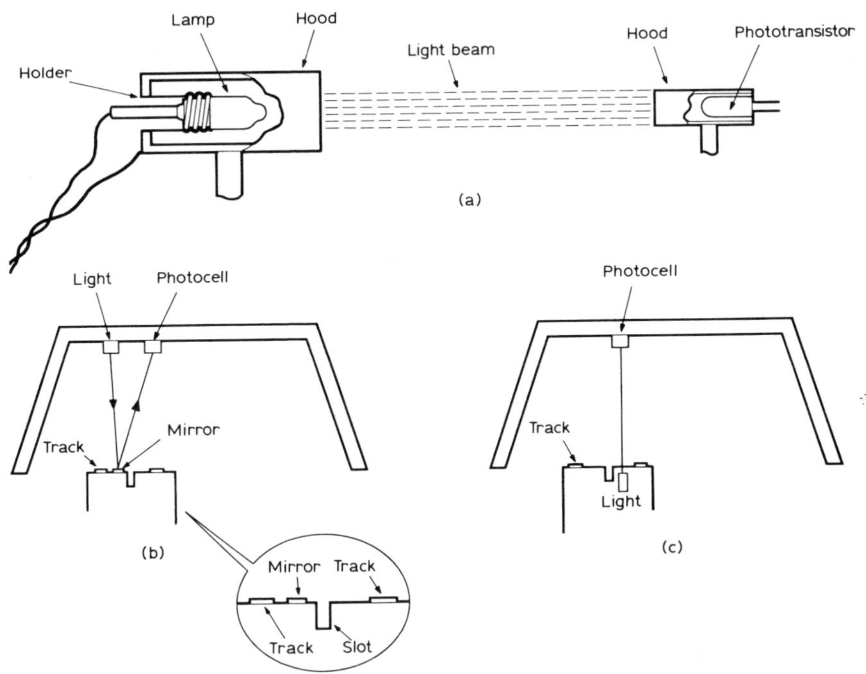

Fig. 5.2 Light beam detector. (a) basic arrangement of components with hoods shown partially cut away, (b) bridge reflection method, (c) using a hole cut in the track to pass the light beam.

circuit to cause a drop in voltage which will turn off a transistor. In this way, a pulse of voltage is generated each time the light beam is interrupted. The voltage pulse at the transistor output can be large enough to operate the counter.

The light beam can be arranged either across-track or from overhead. In the across-track arrangement, the light bulb is arranged at one side of the track, and the photocell at the other; the track here means the track of one car. In the overhead arrangement, the bulb and photocell, each hooded, are located in a bridge over the tracks and a small piece of mirror is placed on one side of the slot. The lamp position is adjusted so that the light reflects from the mirror on to the photocell and will be interrupted by a car. Another possible arrangement uses the bulb under the track and shining through small holes in the track on to photocells mounted in a bridge above the track.

Each arrangement has some drawbacks. The across-track arrangement needs several trackside mountings for bulbs and photocells, and very little space is available if the mountings are not to foul the cars. The overhead arrangement needs careful setting-up and can be upset if the cars reflect back enough light to operate the photocell. The third method is by far the best both for the appearance and the ease of setting-up. Note that any light-beam counting system can be upset by accidental (or deliberate!) interruption of the light-beam by fingers, ties or any other objects.

The bulbs used should be the miniature type with a lens end, as used in miniature torches. The pickup photocells can be photodiodes or phototransistors but of the many types available, only the cheapest need be used. Our circuit suggests the use of an OCP70. Quite often, a bargain bundle of transistors contains a few of the old glass-cased germanium transistors such as OC71. These are usually coated with black paint, to prevent their sensitivity to light interfering with the transistor action. If the paint is scraped off, the transistor will then act as a phototransistor; in fact the OCP70 is simply an OC70 which was not coated with paint.

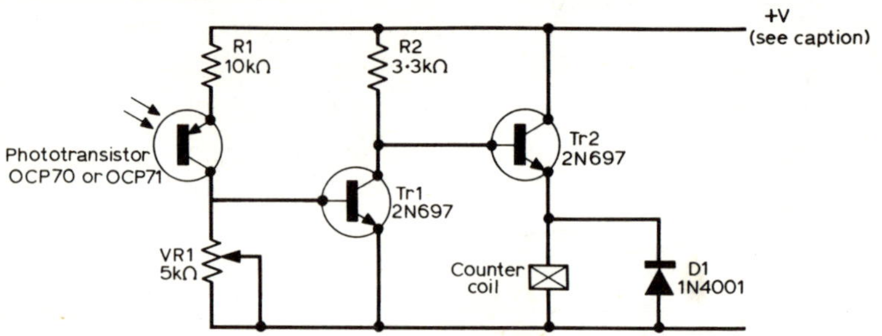

Fig. 5.3 Detector using type OCP70 phototransistor with another two transistors. Most counters operate from 12V, so that a 12V supply line can be used.

Operation of System

With the light hitting the photocell, the resistance of the semiconductor material is low, so that the voltage at the base of Tr1 is high enough to switch the transistor on (Fig. 5.3). A variable resistor here ensures that the voltage is not too high, and can be adjusted so that the amount of light reaching the phototransistor *just* keeps the transistor Tr1 switched on. Because transistor Tr1 is passing current, the voltage at its collector is low, too low to allow transistor Tr2 to switch on. The base (input) of Tr2 is connected directly to the collector of Tr1. No current flows in Tr2, and so no current flows in the coil of the counter.

When the light beam is interrupted, the photodiode or phototransistor changes to high resistance, Tr1 switches off, and its collector voltage rises, switching Tr2 on and operating the counter. In this way, another lap is registered. When the beam hits the transistor again, the action is reversed, and the diode D1 prevents damage to the transistor Tr2 when the current switches off.

The counter can be reset mechanically, so that each race can start with all the counters set to zero. If the light beams cut the starting line, as would be logical, care must be taken that the counters do not register a first lap as the cars leave the start. This is most easily arranged by having a count start/stop switch in series with the connection to the mechanical counter.

Fig. 5.4 Small magnet and pickup coil.

The switch can be set to start just after all the cars have cleared the starting line, and to stop when the first car has crossed the finish. The starting line should not be on a curve, as any car crossing the line sideways will register a count on several lanes, if the overhead system is used, or destroy the laneside detectors if the cross-track method is used.

Magnetic Method

The magnetic method of lap counting uses a small magnet, of the type used for correcting the shape of a TV picture (they are glued round the rim of the cathode-ray tube) which is glued to the car or other object whose laps are to be counted. The trackside detector consists of a coil of wire wound round a soft-iron core, such as the magnet assembly of an old pair of headphones.

When a magnet moves past such a coil, a pulse of voltage is induced in the coil, and the pulses can be amplified and used to operate a counter. The magnet has to pass fairly close to the counter, a few mm at the most, but precise dimensions cannot be given because they depend on the strength of the magnet and the construction of the coil.

The magnetic system has the advantage that it does not register stray fingers or items of clothing, and does not count stationery cars or parts of cars with no magnet attached; but some care has to be taken with the sensitivity of the amplifier to avoid the car's motor magnet operating the count. On the other hand, if all the cars have motors using about the same strength of magnets, it may be an advantage to use the magnet of the motor, since this avoids the need to have any other magnets attached. In this system, minor shunts across the

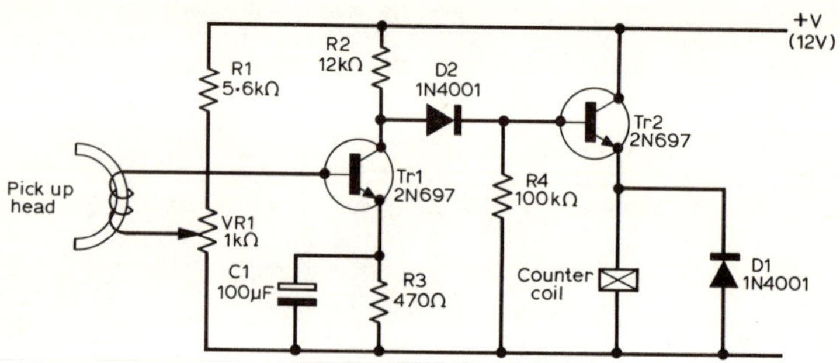

Fig. 5.5 Magnetic detection circuit.

track do not usually cause such errors in counting as happen with the light-beam system.

Amplifier Circuits

The amplifier needed for a magnetic pickup system is shown in Fig. 5.5. One end of the pickup coil is connected to the slider of the potentiometer which acts as a sensitivity control, and the other end to the input of transistor Tr1. A pulse which is negative-going at the input of Tr1 will switch Tr1 off, so that the voltage at the collector of Tr1 rises, turning on Tr2 and so operating the counter. The potentiometer VR1 is set so that satisfactory counting is obtained for the magnet and coil which is used.

A more sensitive version of this circuit, using a type 741 IC is shown in Fig. 5.6. In this circuit, potentiometer VR1 sets the gain of the amplifier so that the system can be made very sensitive, sufficient to detect the movement of the magnet at several cm distance.

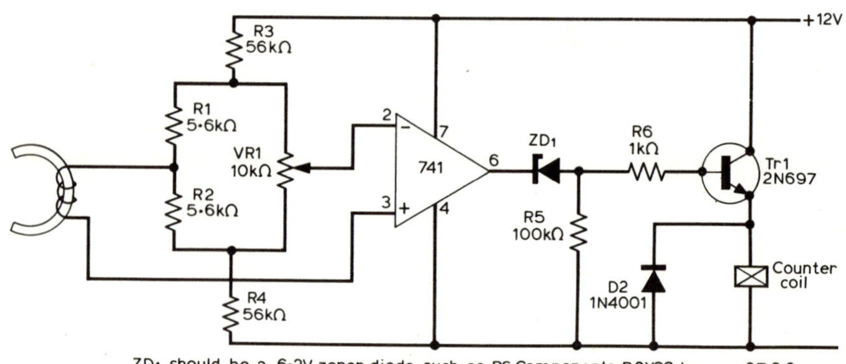

ZD1 should be a 6·2V zener diode, such as RS Components B2Y88 type or 2F 6·2

Fig. 5.6 A more sensitive magnetic pickup circuit than that of Fig.5.5, using a type 741 integrated circuit.

A further advantage of this system is that it can be arranged so that only a magnet connected one way round will operate the counter. For example, we may arrange it so that only a magnet connected with its N pole facing the coil and moving clockwise round the track will produce a negative pulse at the input of Tr1. We cannot very easily predict in advance which arrangement of the magnet or coils will produce this, because it depends on the direction of the winding of the coil.

The easiest check is to set to the minimum sensitivity which will operate the counter in one direction, and then try reversing the magnet. If the magnet does not operate this way round, then the counter will be able to discriminate between two cars on the same track, one carrying a magnet with the N pole pointed to the counter, and the other with a magnet whose S pole is pointed to the counter. This may not always be possible, since some types of magnet or coils may produce a double-peaked pulse for either direction or magnet.

Revolution Counters

Counting techniques can also be used to measure the speed of miniature motors, whatever the power source and, in particular, the speed of miniature diesel or glowplug engines. The circuits which are used to measure the speed of car engines are not suitable, as they depend on picking up the high voltage induced in the spark coil. Digital counters, with a digital readout, can be devised, but at considerable expense, and with the disadvantage of needing a low voltage power supply passing relatively high currents.

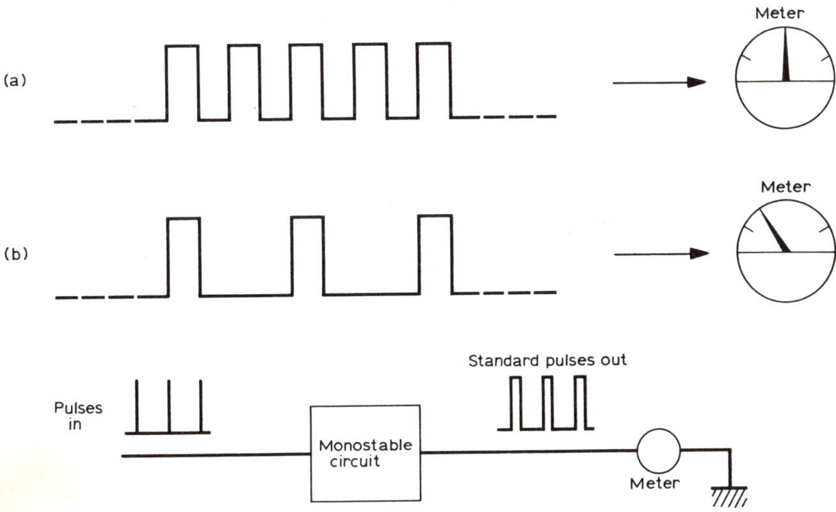

Fig. 5.7 Revolution counting. (a) pulse voltage high for 50% of total time, (b) pulse voltage high for 25% of total time, (c) system arrangment.

A more suitable circuit uses an operational amplifier of the 741 (or similar) type as a monostable, a circuit which provides an output pulse of definite set width and amplitude for each input pulse, If we apply the output of a monostable to a meter, the reading on the meter will depend on the total time in each second the pulse is at high voltage. For example, if we arrange the output pulse to have an on time of 3 milliseconds (3/1000 seconds) and a peak voltage of 6V, and we have a pulse each 6 milliseconds, then the voltage to the meter will be at 6V for 3mS and zero volts for 3mS in each cycle, so that the meter is at 6V for half the time of operation. If the meter reads 6V at full scale, its reading in this example will be 3V, which is half scale.

An input pulse coming every 6mS from a source which gives a pulse at each revolution corresponds to 10 000 r.p.m., so that we have here a system which reads full scale at 20 000 r.p.m., and is linear, meaning that half scale corresponds to 10 000 r.p.m. and quarter-scale to 5 000 r.p.m. We can alter the scale by changing either the pulse amplitude or the sensitivity of the meter, and we can also allow for inputs which give two pulses or more per revolution.

Pulse Generating Circuit

The pulse generating circuit is shown in Fig. 5.8. With no pulse input, the + signal input of the IC amplifier is at the voltage fixed by R1 and R2, and the - signal input is at a voltage just slightly higher than earth voltage, because of diode D2. The output of an amplifier of this type (an Op-amp) is high when the voltage of the + input is higher than that of the - input, so in this case the output voltage is high, keeping the voltage across R1 high. When a negative-going pulse arrives at D1, it reduces the voltage at the + input.

If the pulse is large enough to make the voltage at the + input less than the voltage at the - input, then the output voltage will swing sharply negative. Because of the capacitor C which is connected between the - input and earth, the connection of R3 between the output and the - input does not immedi-

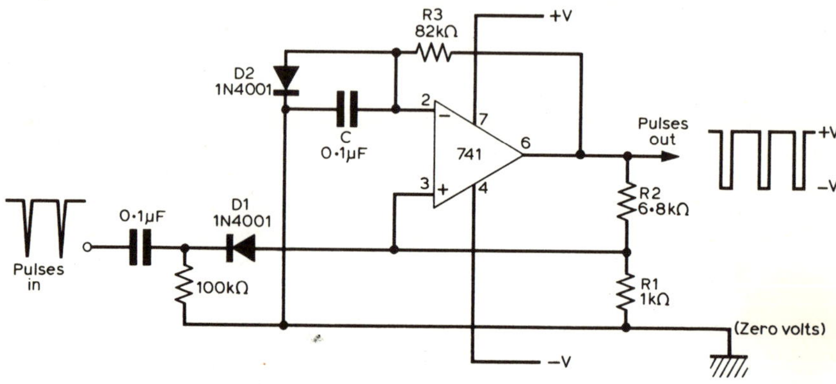

Fig. 5.8 The monostable circuit.

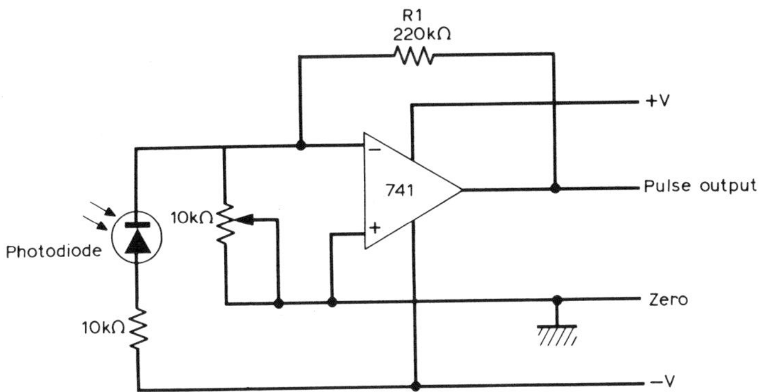

Fig. 5.9 Obtaining the pulses from the light beam.

ately cause the - input to change to a low voltage. The voltage at the - input falls at a rate controlled by the values of C and R3, and when it drops to a value which is less than the value of voltage at the + input the action is reversed and the output voltage becomes positive again.

During the action, the voltage at the + input is always a fraction of the output voltage, set by R1 and R2, and pulses arriving while the pulse is being produced have no affect on the action, because the diode D1 acts to "gate" them out. During the time when the output voltage is low, the voltage at the + input is low, and only a pulse whose voltage is lower still can pass the diode.

Obtaining the Pulses

We can obtain the negative pulses which are needed to activate the counter from interrupting a light-beam, so that the light can be placed on one side of a propeller blade and a photodiode on the other. Photodiodes used in this way must have a fairly rapid response, and some amplification is needed. This can be provided by another op-amp, using the circuit of Fig. 5.9.

In this circuit, light falling on the photodiode causes it to have a low resistance. Because the photodiode is connected between the - input and a negative voltage, light falling on the photodiode will keep the - input at a negative voltage and the output voltage will be high. This high output voltage is connected back to the - input by resistor R1, so that when the light beam is interrupted the voltage at the - input is forced to rise until the output voltage drops enough to lower the voltage at the - input again.

Returning to the measuring part of the circuit, the meter can be a 9V full scale voltmeter, or a 1mA meter fitted with a suitable series resistor. We can adjust the meter reading for any given r.p.m. by altering the value of the series resistor, or by altering the width of the pulse. If only slow revolution rates are to be measured, wider pulses can be used, but if revolution rates in excess of 20 000 r.p.m. (or rather 20 000 pulses per minute) are to be used, then shorter pulses should be used.

Since speeds in excess of 10 000 r.p.m. can be obtained from most mini-

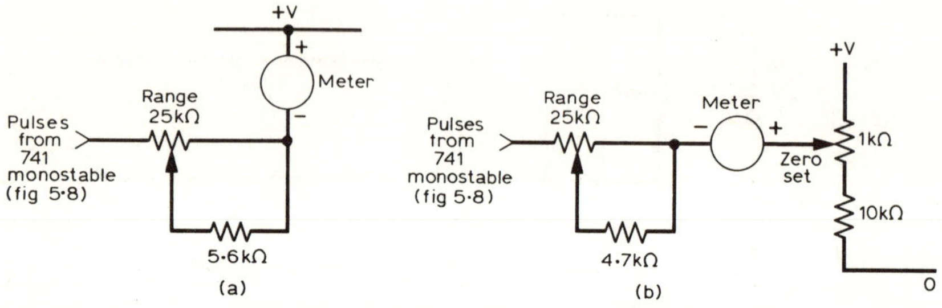

Fig. 5.10 The meter circuit. (a) basic circuit, (b) a circuit which may be needed if the meter needs a zero-setting control. At some supply voltages, the output of the 741 may not equal the +V level between pulses, and the meter will have a voltage across it, causing a reading with no pulses in. The 1kΩ variable resistor enables the meter to be set to zero with no pulses in.

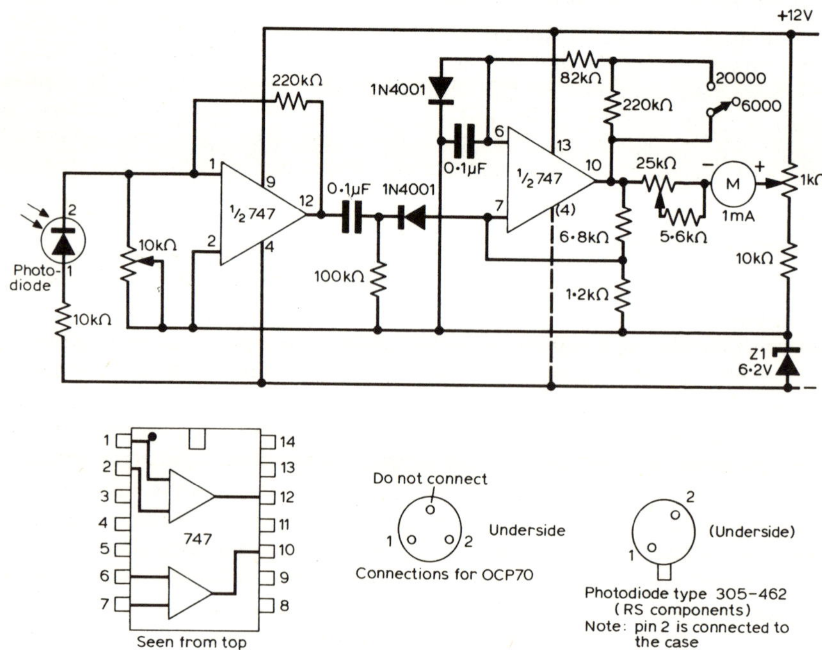

Fig. 5.11 Complete circuit, using a type 747 IC. For this version, only a single 12V supply is needed. The photodiode shown is obtainable from RS Components, but photodiodes from the TIL series, or a phototransistor, can be used. The connections for a type OCP70 are also shown.

ature glowplug engines, and a two bladed propeller will result in the number of pulses per minute being twice the number of r.p.m., a shorter pulse, perhaps 2mS, should be used for aircraft work, and the sensitivity set by altering the meter resistor.

Op-amp Circuit

A considerable simplification in the circuitry is obtained by using a type 747 dual op-amp as shown in Fig. 5.11, which shows the complete circuit for ranges of 6 000 r.p.m. and 20 000 r.p.m., assuming a two-blade propeller. Calibration of the meter can be carried out by exposing the photodiode to a mains-operated fluorescent light. These flicker at 100Hz (6 000 pulses per minute.)

Some adjustment of the brightness may be needed, using photographic negatives in front of the photodiode to cut down the amount of light, and no other lights should be switched on. The reading obtained should be adjusted so that the meter reads half-scale on the 6 000 r.p.m. range, assuming two pulses per revolution. This will appear as a reading of 15% along the scale on the 20 000 r.p.m. range.

If the unit is to be used for measuring a variety of different speeds, with two or three-blade propellers, it is probably better to calibrate the unit in terms of pulses per second. In this case, the reading on the fluorescent light will be adjusted to be full scale on the 6 000 pulses per minute range and 30% along the 20 000 pulses per minute range. The speed will then be obtained by measuring the number of pulses per minute and dividing by the number of propeller blades or other interruptions per revolution.

Using Sound Impulses

Sound impulses can sometimes be more useful for either lap counting or for revolution counters. For control-line model aircraft, cutting a light beam is not particularly easy, since the transmitter and receiver have to be a fair distance apart, and one above the other if the aircraft is to cut the beam. Any movement of the pilot from a central point will take the path of the plane away from the counter, so that the scheme is rather unsatisfactory. In this case, the sound output of the motor provides a better method of operating a counter.

If a microphone is fixed up at one point in the circle defined by the control lines, the sound reaching the microphone will be very much more intense when the aircraft is close to the microphone than when it is distant. Sound energy obeys (in the absence of echos) an inverse-square law, meaning that doubling the distance between a source and a detector makes the intensity one quarter of its previous value, and tripling the distance makes the intensity 1/9 of its former value.

If we place a microphone so that the model at its point of nearest approach is about 4 feet from the microphone, then on 50′ lines, the furthest point is 54′ away, and the sound intensities at these distances should be in the ratio $54^2 : 4^2$, which is about 182:1, large enough to be easily distinguished. Some care has to be taken about wind direction; it is better to arrange the microphone so that the direction of the wind is from the microphone towards the centre of the circle.

Pickup Circuit

The circuit for the pickup consists of a microphone, which can be a cheap dynamic type, feeding into a high gain amplifier. This will have an output at the frequency of the motor; for a motor turning at 9 000 r.p.m. the frequency will be 9 000/ 60, which is 150Hz. This output of 150Hz will rise and fall in amplitude as the model flies round the circle and is therefore alternately near to and distant from the microphone.

The shape of the wave produced is described as amplitude modulated, and we cannot use it directly to operate a counter because of the 150Hz component which would cause the counter to try to operate at this frequency. The

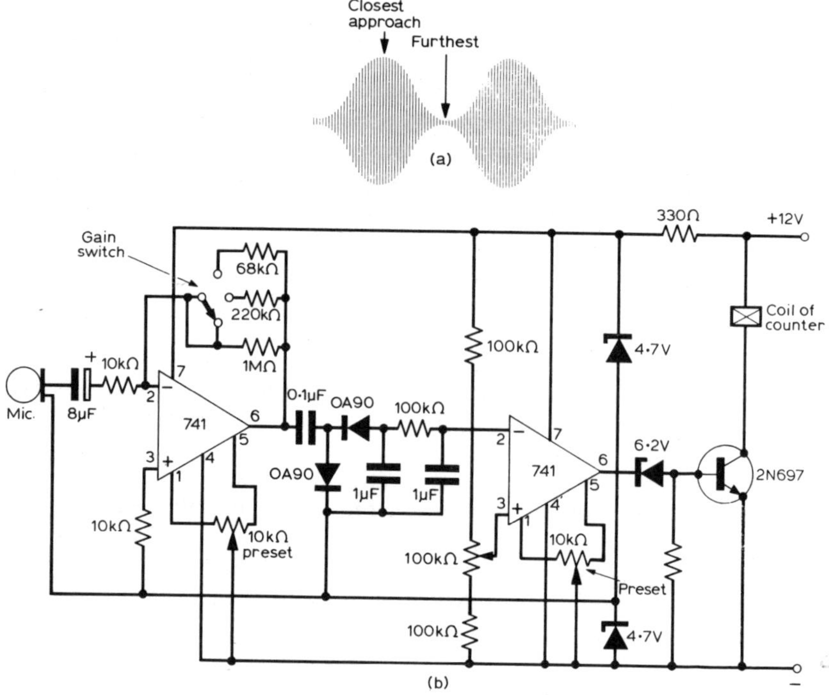

Fig. 5.12 Using sound triggering. (a) sound wave pattern: the high frequency is the motor noise and the low frequency represents the change in loudness caused by the model being alternately near to, and distant from, the microphone, (b) circuit for lap counting using sound triggering: the counter coil is assumed to operate from a 12V supply.

wave is therefore passed through a diode and a filter, which removes the higher frequency and leaves us with a wave which rises and falls in amplitude at the frequency with which the plane laps the circle.

This resulting wave can be used to trigger a monostable, which generates an output pulse each time the voltage of the wave crosses a level set by a "trigger level" control. The number of pulses from the monostable can then be used in a counter of the type used in Fig. 5.3 or 5.5. The complete circuit, with adjustments for amplification and trigger level, is shown in Fig. 5.12b.

If we want to use the sound of the motor to operate the revolution counter we must make sure that the pulses from the microphone are clearly separated from each other. This time, of course, we are aiming at using the "high" frequency of the motor. After amplification, the pulses are put through a differentiating circuit which sharpens up the pulses. Ideally, the circuit should be adjustable so that the sharpest possible pulse can be obtained for the motor frequency which is in use, but the values given in the circuit present a reasonable compromise for most model diesel and glowplug motors.

The differentiated pulses are then used to switch the monostable circuit of Fig. 5.13 and so provide the revolution counter reading. As the motors used are mainly two-strokes with one exhaust pip per revolution, the counter will read r.p.m. directly. Owners of two-lobe Wankel motors will have to make their own arrangements!

Time and Speed Measurements

Of all the quantities which have to be measured in competitive work, high speeds and short times present the greatest difficulties. Average speed is often

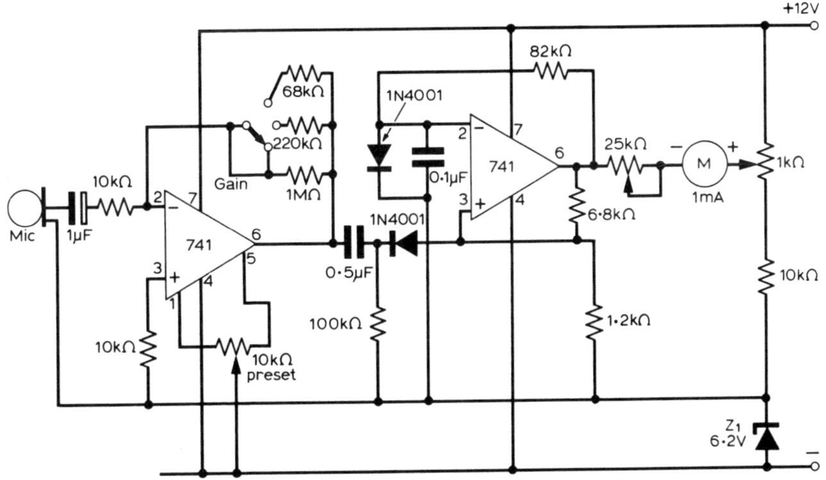

Fig. 5.13 Circuit for a revolution counter using engine sound.

measured by timing a model over a measured distance using a stop-watch. Because of the "thinking-time" (human reaction time) which every user of a stopwatch will add to or subtract from a reading, these times are reasonably precise only when the time is fairly long, a minute or more, and there is no way of using a stopwatch with a human operator switching it on and off which will give reliable timing of times of the order of a second or less, no matter how good the watch.

Electronics methods can be used to measure very short times and very high speeds with much more confidence, so that the maximum speed along a straight or nearly-straight path can be measured over distances little more than the length of the model itself. In an electronic system, the switching on and off must be carried out electronically, so eliminating the human factor which makes the use of conventional stop-watches unreliable, and the counting of time is also done electronically.

The principle used is that of generating a waveform of a precise frequency, such as 1kHz. The clock has two inputs, a switch-on and a switch-off. A pulse to the switch-on circuit starts the generator, and a pulse to the switch-off circuit stops the wave. The number of cycles of the wave are counted, giving the time in terms of the number of waves. The accuracy is plus or minus one wave. For timing a model, the switch on and switch off pulses would have to be provided by such methods as cutting light beams.

The construction of a digital stopwatch is beyond the scope of this book, which is concerned with the type of projects suitable for a comparative newcomer to electronics. An excellent design for the more venturesome constructor with some experience is detailed in *Electronics Today International* for January 1974, and reprinted in the booklet, *E.T.I. Top Projects No. 2*, obtainable from bookshops or from E.T.I. Specials, E.T.I. Magazines, 36 Ebury St., London SW1W 0LW (the price at the time of writing was 75p + 15p post). As for most of the projects published by this magazine, printed circuit boards are available ready made from Crofton Electronics Ltd., 124 Colne Road, Twickenham, Middx.

CHAPTER SIX

LOGIC CIRCUITS

IN THE SIGNALLING SYSTEMS traditionally used in railways, safety was ensured by interlocking controls so that, as far as could be arranged, human error was avoided. One simple example of this interlocking is that the lever which controls signals cannot be pulled to the "road clear" position until all the points in the path of the train have been set in the correct position. In the old signal boxes, these interlocks were mechanical, operated by locking bars set across notches in the operating rods. In the more modern boxes, these interlocks are electrically operated, but model control systems very seldom attempt any kind of interlocking system.

This reluctance to undertake interlocking is fairly easy to understand, because to scale down either a mechanical or an electromechanical interlock system would be an extremely difficult task. It is now possible, however, to use full interlocking in a model railway layout by making use of electronic logic circuits, which are also coming into use in the interlocking systems used on the full size systems. These logic circuits make use of the modern techniques of integrated circuits to pack a very large amount of circuitry into a very small space, and a few basic circuits can be used to perform very complex interlocks by suitable interconnection. In this chapter we shall examine what these circuits do and how they can be used for modelling purposes.

Basic Logic Actions

The basic logic actions are AND, OR, and NOT. The circuit called the AND gate has several inputs and one or two outputs, generally one only. These will be a high output voltage (symbolised as 1) when *all* of the inputs are high (all at level 1), so that for a two-input, one output gate there will be an output of 1 if input (a) *and* input (b) are both high (at 1). This is the equivalent of an interlock, because there is no output unless both of the inputs are present. We could imagine a two-input AND gate as part of a signal system, as in Fig. 6.1a, where the signal-on switch has no effect unless the points are also energised at the same time.

The OR circuit will give an output of 1 if any one of its inputs is at 1. For a two-input OR gate, there is a 1 output if (a) *or* (b) is at 1. We can imagine

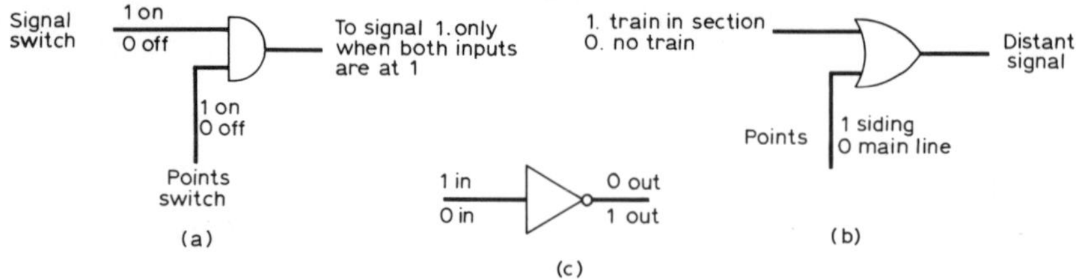

Fig. 6.1 Gates. (a) AND gate, (b) OR gate, (c) NOT circuit.

this circuit used to keep a distant signal red when there is a train already in the section *or* when the points are set to a siding. The NOT circuit gives an output which is the reverse of its input, and there is only one input and one output. For a zero (0) signal in, there is a high (1) signal out, and for a high (1) signal in there is a zero (0) signal out.

In practice, the NOT circuit is usually combined with one of the other two as well as being available separately. The most commonly used logic gate is the NAND gate, which combines NOT with AND, so that the output is low (0) if all the inputs are high (1) and the output is high (1) for any other combination of inputs.

Because so many words are needed to describe these simple actions, a **truth-table** is a more convenient way of showing the action of any logic element. The truth table shows what the output, or outputs, will be for any combination of inputs, and is more convenient to refer to when we are tracing the action of several gates. Truth tables for AND, OR, NOT, NAND and NOR gates are shown in Fig. 6.2, along with the symbols commonly used for these gates.

TTL and CMOS

The amount of progress which has been made in the design of electronic logic systems has been large and very rapid, so that system after system has become available and then obsolete. At the moment of writing, the older TTL system, itself comparatively modern, is being replaced for many purposes by the more modern CMOS system. The distinction, as far as the non-specialist is concerned, is that the TTL systems are easier to construct and use, though the CMOS system has several advantages.

The overwhelming drawback of the CMOS system for any constructor who is not a dedicated electronics fan is that the circuits are easily damaged by the electrostatic voltages which are built up on the human body, or on unearthed soldering irons, so that the circuits are very vulnerable to damage by handling until they are safely soldered into circuit. The problem is overcome to some extent (and is not too serious if some care is taken) by keeping all the pins of the circuit "shorted" together by a pad of conducting foam plastic which is removed only after the circuit has been soldered into place in the printed cir-

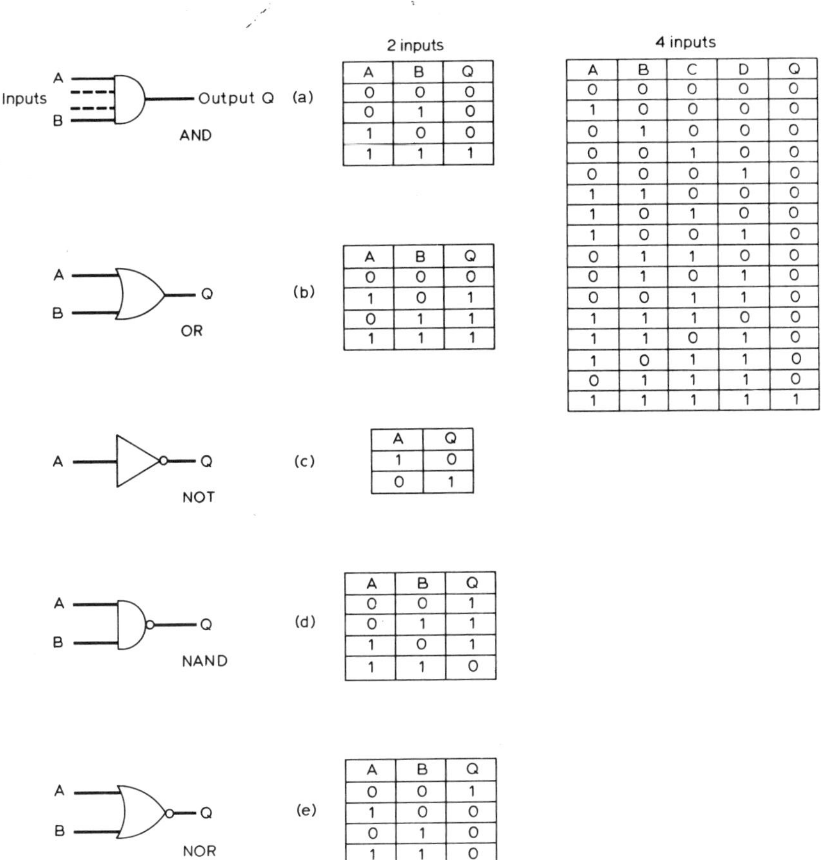

(a) 2 inputs

A	B	Q
0	0	0
0	1	0
1	0	0
1	1	1

4 inputs

A	B	C	D	Q
0	0	0	0	0
1	0	0	0	0
0	1	0	0	0
0	0	1	0	0
0	0	0	1	0
1	1	0	0	0
1	0	1	0	0
1	0	0	1	0
0	1	1	0	0
0	1	0	1	0
0	0	1	1	0
1	1	1	0	0
1	1	0	1	0
1	0	1	1	0
0	1	1	1	0
1	1	1	1	1

(b)

A	B	Q
0	0	0
1	0	1
0	1	1
1	1	1

(c)

A	Q
1	0
0	1

(d)

A	B	Q
0	0	1
0	1	1
1	0	1
1	1	0

(e)

A	B	Q
0	0	1
1	0	0
0	1	0
1	1	0

Fig. 6.2 Truth tables and symbols. (a) AND gate, (b) OR gate, (c) NOT circuit, (d) NAND gate, (e) NOR gate. The symbols shown are those used by American manufacturers; British Standard symbols exist, but are not so commonly used.

cuit board. The CMOS circuits are soldered in only after all the other circuit components have been mounted, so that each pin of the CMOS integrated circuit has a path through a resistor to a common earth.

The alternative TTL circuits are not so easily damaged in the construction stage, and need protection only against excessive voltages during operation. Since TTL circuits have been made in very large numbers by several major manufacturers and are available cheaply on the "surplus" market, it seems likely that they will be easily available for many years to come, and so they have been specified for the logic circuits used in this book. Like all electronic integrated circuits, TTL (the letters are an abbreviation of Transistor-Transistor Logic) circuits have to operate under specified conditions. The standard supply voltage is 5V, which is not readily available from a battery, though a fresh 4·5V dry battery can be used to operate a few circuits. The current consumption of TTL circuits is comparatively high, so that we would normally use mains power packs to provide the 5V output, and suitable supplies are easily built (see Chapter 8).

TTL Principles

Naturally, the logic for each different railway layout must differ, so that we can only show operating principles here, but a knowledge of these principles should enable the reader to convert any layout to electronic logic. For large layouts there is also the tantalising possibility of approaching 'real-life' action

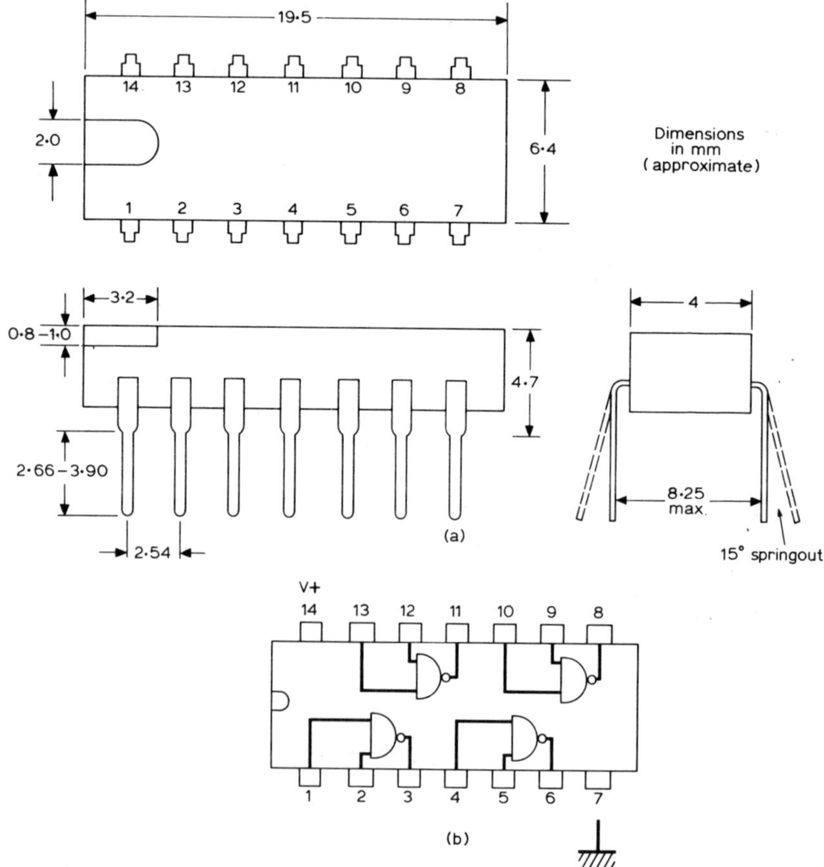

Fig. 6.3 TTL gates. (a) outline of package, 14-pin, (b) connections for NAND gate type 7400.

Fig. 6.4 TTL integrated circuits.

in terms of having one-switch route controls, so that the operation of one switch will actuate all the points, energise the track and set the signals for a complete route. There is even the possibility of programming a complete timetable onto a cassette recorder, so that replaying the cassette into the control centre will operate the assembly of trains, route selection, and running to timetable.

Since all the logic systems operate from a 5V supply, we shall either have to operate signals and points from 5V or, more suitably, use transistors between the logic systems and the points, signals etc. In the language of computer designers, we need to use transistors for interfacing.

The family of TTL circuits in the 74 series, in which each reference number starts with 74, are obtained in the form of dual-in-line flatpacks (Fig. 6.3), which are small flat blocks about 8mm wide by 20-22mm long with 14 or 16 pins. Because so many pins are available in the standard pack, and because practically any number of circuits can be made in a pack for about the same cost, it is usual to have several gate circuits in one pack. The type 7400N, for example, contains four two-input NAND gates; the 7410N contains three three-input gates, and the 7420N two four-input gates.

Single Interlock

For a single straightforward signal-points interlock, one of the gates of a 7400N could be used. The signal and points switches are simple on/off switches operating from a 5V supply, and the inputs to the gate are taken from the switch contacts as shown in Fig. 6.5. As a result, the gate is at zero output when both switches are ON, but at high output when one or more switches are OFF.

The next step is the operation of the signal from the change in gate output from high to low. If the signals are being built in a new system, considerable savings can be made by using miniature LEDs as light signals, since these can be

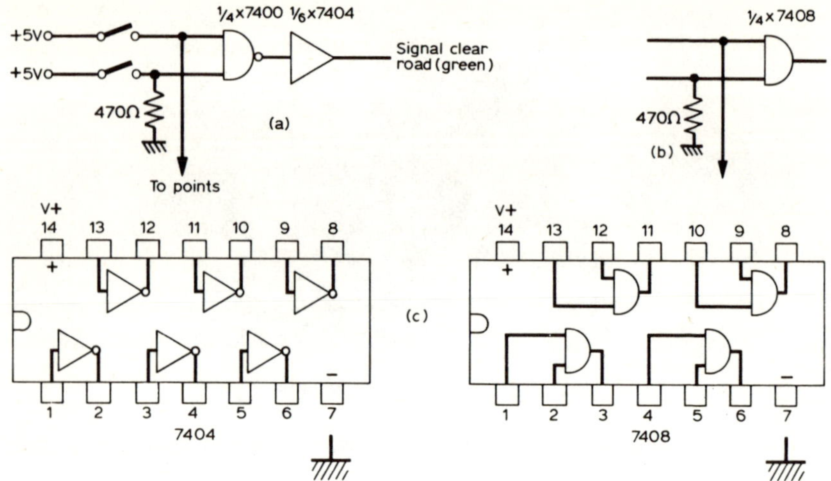

Fig. 6.5 Simple signal/points interlock. (a) using NAND gate and inverter, (b) using AND gate, (c) 7404 and 7408 pin diagrams, as viewed from the top. The cost of the NAND gate and inverter is very little more than that of the AND gate alone, so that the use of the NAND gate is a better proposition, since inversion is usually needed anyway.

operated directly from TTL voltage supplies. Either a transistor or one of the inverters in a multiple NOT gate (such as the 7404N) can be used to change the output of the NAND gate to the polarity required for the signals, as in Fig. 6.5a.

If filament lamps or solenoid-operated lever signals are used in an existing layout, different voltage and current levels may be needed, and transistor operation is needed to make the logic system compatible with the control systems. This type of circuit, then, will ensure that the setting of the points controls the operation of the signals, so that the signal cannot be set to green unless the points have first been correctly set.

Two-point Interlock

We can easily extend this system to allow a green signal only if two or more points have been correctly set. Fig. 6.6 shows the arrangement needed if two sets of points have to be switched to allow a signal to set to green, and also to ensure that the red signal shows when the green is off, and the other way round. An inverter is used between the red and green control lines, so that the red signal is controlled by the green, and the two can never light at the same time.

The logic of this circuit is that the signal can be switched to green only if both points are correctly set and the signal switch set to green. At all other times, the signal shows red. A small change in the circuit layout can cater for

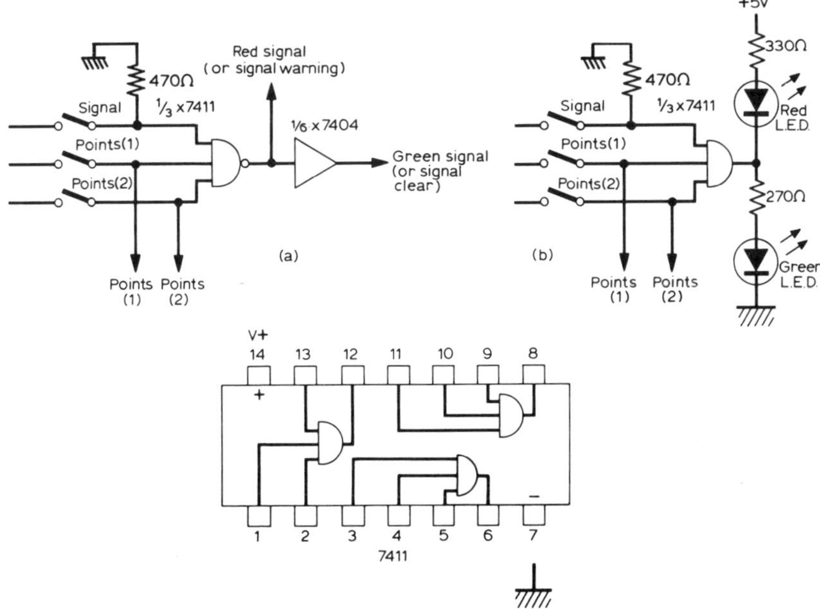

Fig. 6.6 Interlock of signal with two sets of points. (a) using 7411 and 7404 for driving signals, (b) driving LED signals directly, using a 7411.

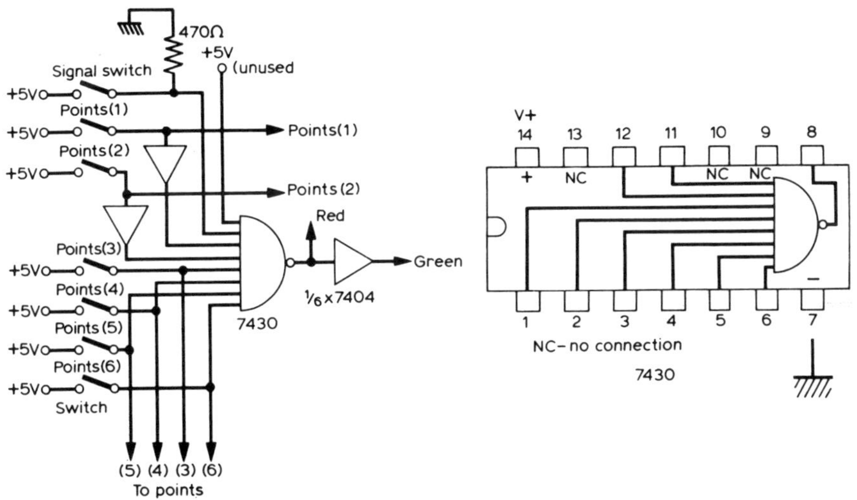

Fig. 6.7 An example of an elaborate interlock where seven inputs control the output.

the case where one of the points has to be energised and the other not, for the road to be clear and a green signal acceptable.

Fig. 6.7 shows a more elaborate layout in which some points need to be energised and some others need to be left unenergised. The voltages from the points switches are taken to the gate either directly, when the point has to be energised to clear the road, or through inverters in the case of points which must remain un-energised. Note that when any inputs to a gate are not used, they must be taken to the +5V line, otherwise the gate cannot ever switch over. Once again, an inverter must be used to switch the red light on when the green light is off, and the red on when the green is off.

Complete Route

We can extend the principle further. Suppose a complete route is to be cleared, setting points and signals ahead of a train in each section of the track, and also cutting the power to the rails if any section is not correctly set. Suppose the layout is divided into three sections. The power to each section can be switched on and off by a thyristor (see Chapter 7) or by a relay operated by a transistor; in each case controlled by the TTL logic circuits, and the points and signals in each section can also be switched as described earlier in this chapter.

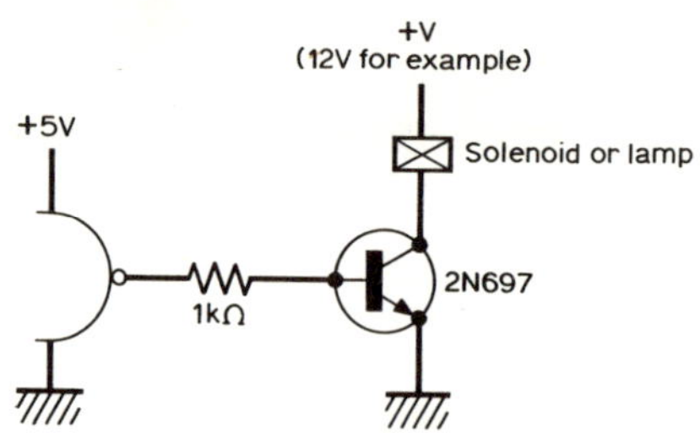

Fig. 6.8 Operating solenoids and filament lamps from TTL logic circuits.

We can now use a single route switch to provide an input to each of the gates which will clear the whole route at one operation, energising all the points which must be changed over, leaving the rest unenergised setting all the signals, and energising the rails in all the sections which form part of the route.

Programming Signal/Points Systems

Imagine that we have several sections of track which can be set, as outlined above, by applying a 1 signal to a gate and released by applying a 0. We can also control the rail voltage by similar signals, so that the whole layout can be controlled to different routes by applying 1 or 0 voltages to different gates. Suppose that we have ten such sections. We can switch any of these ten sections

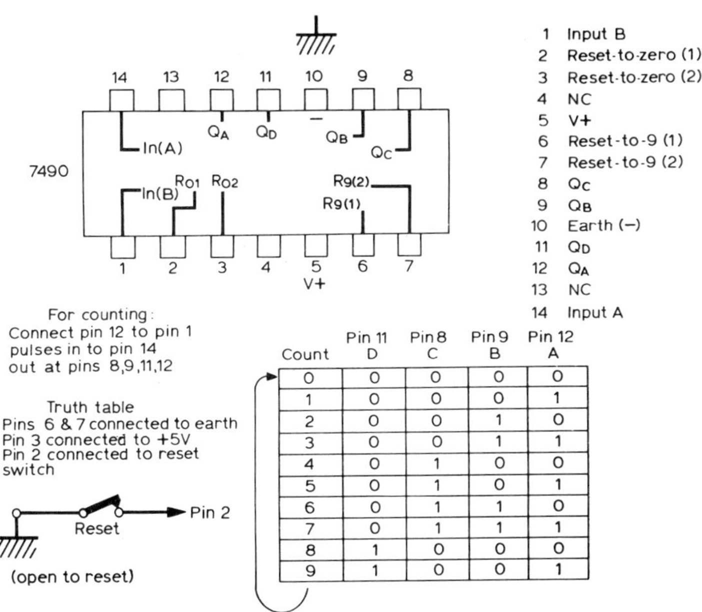

Count	Pin 11 D	Pin 8 C	Pin 9 B	Pin 12 A
0	0	0	0	0
1	0	0	0	1
2	0	0	1	0
3	0	0	1	1
4	0	1	0	0
5	0	1	0	1
6	0	1	1	0
7	0	1	1	1
8	1	0	0	0
9	1	0	0	1

Fig. 6.9 Decade counter – pin diagram and truth table for 7490.

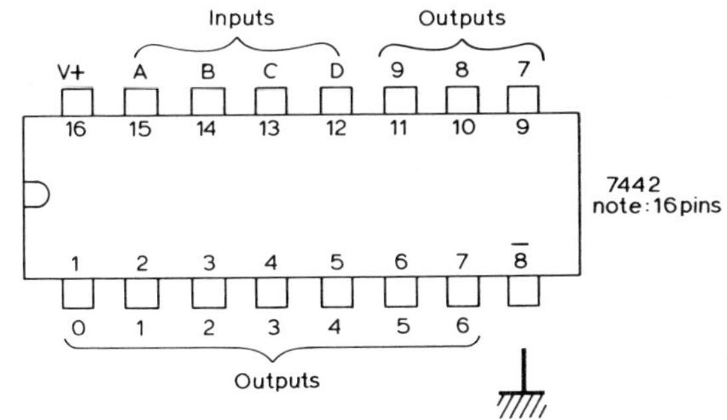

Inputs				Outputs									
D	C	B	A	0	1	2	3	4	5	6	7	8	9
0	0	0	0	0	1	1	1	1	1	1	1	1	1
0	0	0	1	1	0	1	1	1	1	1	1	1	1
0	0	1	0	1	1	0	1	1	1	1	1	1	1
0	0	1	1	1	1	1	0	1	1	1	1	1	1
0	1	0	0	1	1	1	1	0	1	1	1	1	1
0	1	0	1	1	1	1	1	1	0	1	1	1	1
0	1	1	0	1	1	1	1	1	1	0	1	1	1
0	1	1	1	1	1	1	1	1	1	1	0	1	1
1	0	0	0	1	1	1	1	1	1	1	1	0	1
1	0	0	1	1	1	1	1	1	1	1	1	1	0
1	0	1	0	1	1	1	1	1	1	1	1	1	1
All other combinations				1	1	1	1	1	1	1	1	1	1

Fig. 6.10 BCD-to-decimal decoder 7442.

on or off by a single set of pulses applied to a counter, so that section 1 is on, then off, then section 5 on, then off, and so on in a programme which consists of a series of pulses. Since the programme of pulses can be stored in the form of signals on an ordinary cassette recorder, it is possible to have a system which will automatically timetable the routing, leaving the operator to sort out the train control.

The system which can be used for this type of control is a decade counter and BCD-to-decimal decoder. The decade counter takes an input consisting of a set of pulses at an input, and produces a set of outputs on four output lines – A, B, C, and D. The BCD (Binary-coded decimal) decoder accepts the four outputs from the decimal counter at its four inputs, and converts these to decimal form, with 9 outputs labelled 0 to 9. The number of pulses into the decimal counter is thus converted to a voltage on the appropriate output.

For example, five pulses into the decimal counter will produce a 0 output (low) on the pin labelled '5', and eight pulses on the input will produce a low voltage on the output pin labelled '8'. In this way, then, we can obtain an '0' output on any of ten pins for a set of pulses in, and the pulses can be from a

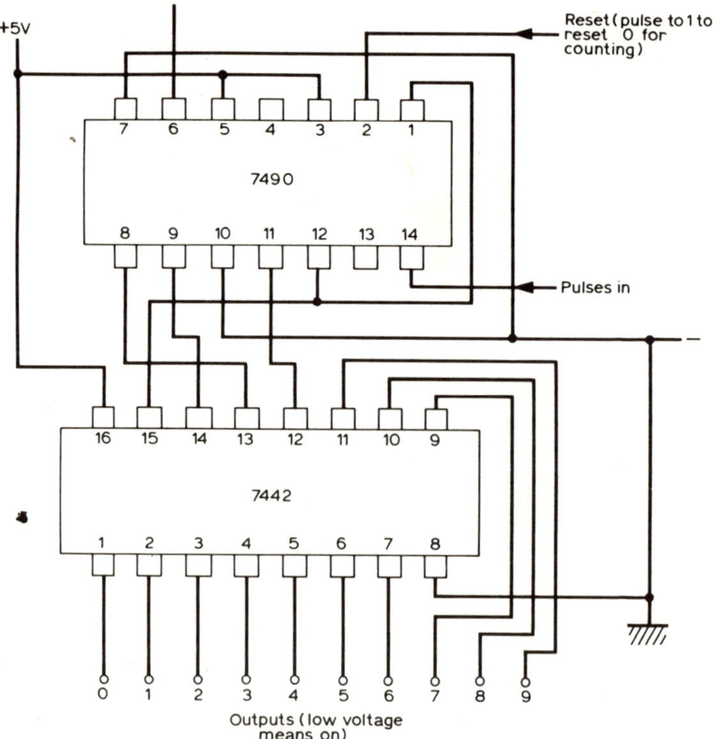

Fig. 6.11 Interconnection of 7490 and 7442 to form a pulse counting unit. Note that the outputs are zeros, so that inverters will be needed if positive outputs are to be used.

signal generator or any other source. This is, of course, an elaborate control system which would be used only if we were attempting automatic single-handed control of a large system.

Before each group of pulses appears at the input, the counter must be reset so that the new set of pulses is not simply added to the previous number. This can be done by a separate resetting pulse to each part of the counter, but it is easier to arrange the counter so that each 10th pulse resets, and we then have to use numbers of pulses which will reset and start the count again.

For example, if we have to code the sequence 9-2-7-1-4, we send in 9 pulses, so switching on the section controlled by the 9th output. After some time, when the movement in this section is complete, the next command consists of three pulses, one to add to the nine to reset the count, and the other two causing an output on the 2nd output pin. To reach the seven we then need 5 more pulses, and the 1 is reached by another 4 pulses (10 − 7 + 1). Finally, another three pulses give the output on pin 4 which ends the sequence.

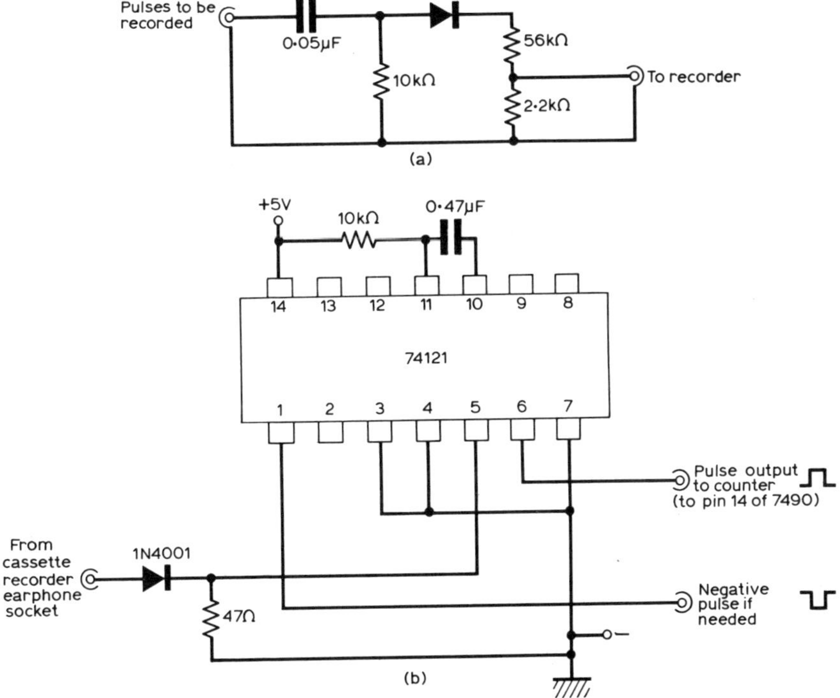

Fig. 6.12 Recording pulses on a cassette tape recorder. (a) input circuit; the socket marked 'to recorder' should be connected to the 'radio in' or 'line in' socket of the recorder, not to the microphone input, (b) output circuit; the monostable type 74121 reshapes the distorted pulses from the recorder into fresh pulses about 3mS long.

Pulses from Tape Recorder

To obtain operating pulses from a cassette recorder, the output from the recorder taken from the earphone socket through a miniature jack plug, should be

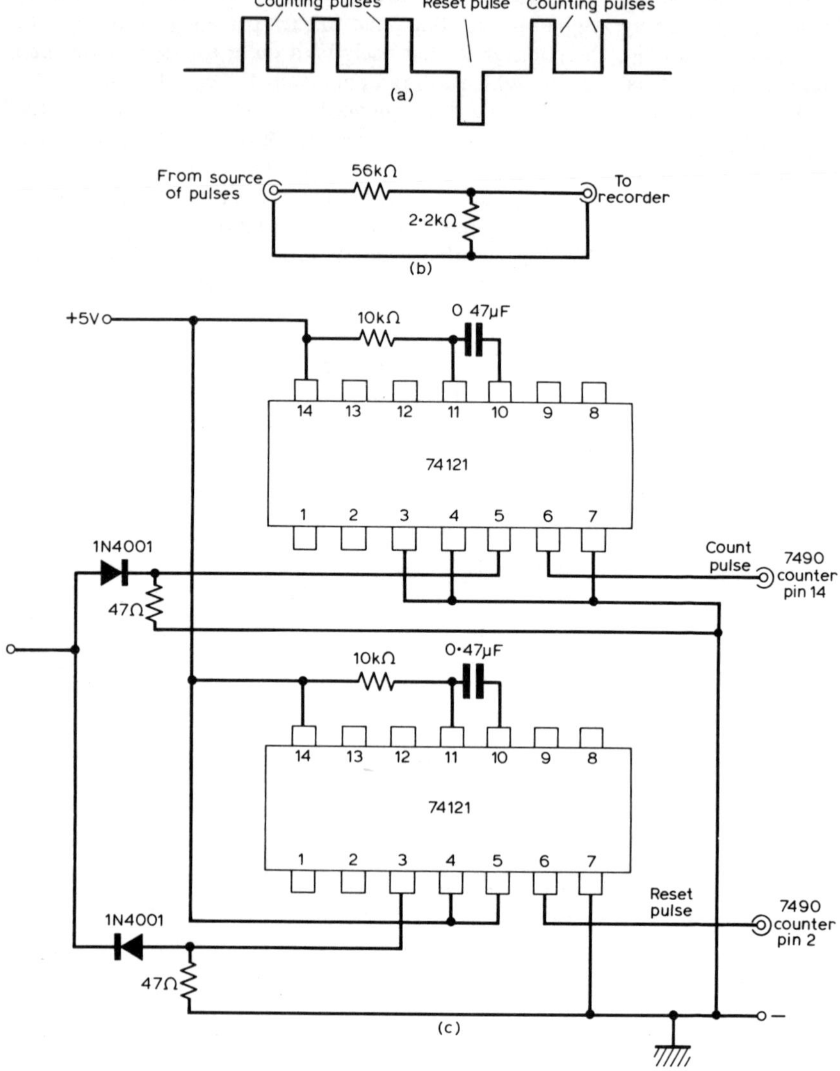

Fig. 6.13 System of using separate reset pulses. (a) using positive counting pulses and negative reset pulses, (b) recording circuit, (c) replay circuit. Note that the two 74121 circuits are differently connected, since the lower one on the diagram has to operate from a negative input pulse.

fed into the circuit shown in Fig. 6.12 so that the distortions produced by the recording process and replay are removed, and the pulses sharpened up for use with the TTL circuits. Note that this comparatively simple counting system has the disadvantage that it is not self-correcting. A poorly shaped or missing pulse will cause the whole sequence to go wrong.

If a separate reset pulse can be used between each sequence, a missing pulse will cause an error in only one section; this is possible if the reset pulses can be separated from the programme pulses. A possible method is outlined in Fig. 6.13 in which the reset pulses are of opposite polarity to the programme pulses and are separated by a more elaborate replay decoder, with the reset pulses then taken to the reset pins of the decimal counter and decoder.

The pulses must be generated somehow, and a suitable pulse generator is shown in Fig. 6.14, for the simpler arrangement which uses no separate reset pulse. The generator consists of a Morse key and pulse shaping circuit arranged to give a pulse 3mS long at each tap of the key. The single output from this generator is taken to two sockets, one for connection to the "radio" recording socket of a tape or cassette recorder, and the other for connection to the track

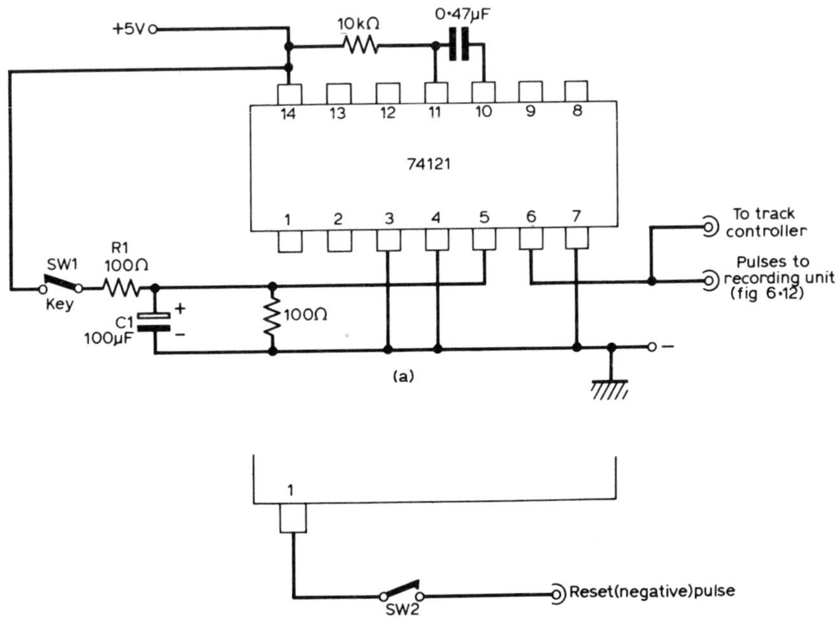

Fig. 6.14 Generating the pulses. A simple Morse key does not generate pulses of steep rise time or constant width. The addition of the 74121 circuit (a) gives clean pulses about 3mS wide, and the use of R1 C1 at the input prevents contact bounce from affecting the count. To use the circuit, the key is tapped for the number of pulses required. (b) to add a reset pulse, the negative pulse from pin 1 is taken through Sw2; for reset, Sw2 is pressed, then Sw1.

controller, so that the effect of the signals can be seen as they are recorded.

A slightly more elaborate system which records a separate reset pulse is shown in Fig. 6.14b. The reset pulse is operated by an additional switch; normally a reset pulse would be used just before a new command, but for some applications a reset may be used just after a command so that there is a period in which there is no activity on the track.

Pulses and Radio Control

The use of pulses to switch different circuits need not be confined to rail track systems. Though radio control is outside the scope of this book, the logic pulses which have been described can be transmitted and received by standard radio control units, and the same coding and decoding methods used, though the reset pulse has to be rather differently treated. The size and weight of a ten-channel decoder make this method more suitable for boats than for aircraft, and proportional control can be achieved by making any selected channel switch on and off by sending number pulses followed by reset pulses.

If the number pulses are not followed too closely by the reset, the average voltage at the output pin will be high; if the number pulses are followed in rapid succession by the reset pulse, with a gap before the next number pulses, then the average voltage on the output pin will be low, so that a mark-space form of proportional control is possible. The reset pulse would have to be a longer pulse, since pulses of opposite polarity cannot easily be transmitted. Fig. 6.15 shows an outline of the circuits which are needed for coding and decoding. Note that this 10-channel system needs only a single radio frequency.

By adding another stage of counting to both coder and decoder, 100 separate commands can be conveyed. Such an amount is seldom needed on radio control equipment, and we would have to use higher pulse repetition

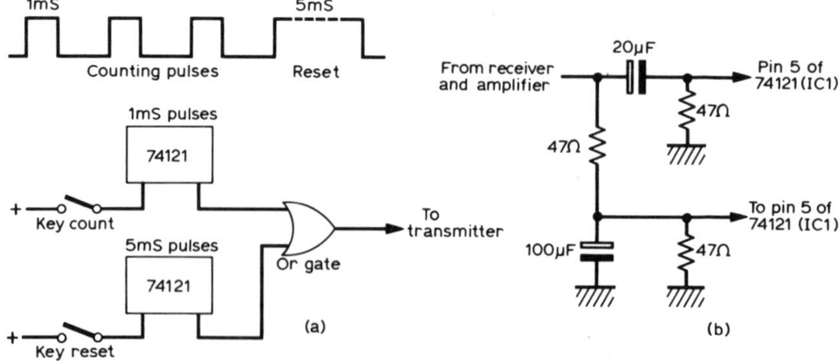

Fig. 6.15 Pulses for a radio control system. (a) pulses used and simple pulse generator, (b) separating the pulses at the receiver. This is possible only if the output is taken from an amplifier capable of passing 5-10mA into these circuits.

rates to avoid serious time delays as the pulses were counted. Rail layouts are more likely to need large numbers of controls, and are less troubled by time delays, for even 100 pulses at a repetition rate of 500 per second will take only 0·2 seconds to count. The reason for using such a comparatively low frequency is to enable a cassette recorder to deal with the pulses, since the amplitude and shape of pulses at high repetition rates will be severely distorted by the simple recordings and replay systems of cheap cassette recorders.

CHAPTER SEVEN

MOTOR SPEED CONTROL CIRCUITS

MOTOR SPEED CONTROL has been mentioned briefly in Chapter 2, but the control of motor speed is such an important topic, not only from the aspect of controlling the speed of miniature electric motors but also the control of the larger mains motors used for drills and other constructional tools, that motor speed control deserves a chapter to itself. Speed control can mean setting a motor to a constant speed, irrespective of load, as well as being able to alter the speed of a motor. To understand how modern electronic systems of speed control operate, we need to know rather more about small electric motors.

P.M. Motors

The smallest type of electric motors are usually permanent magnet d.c. motors.

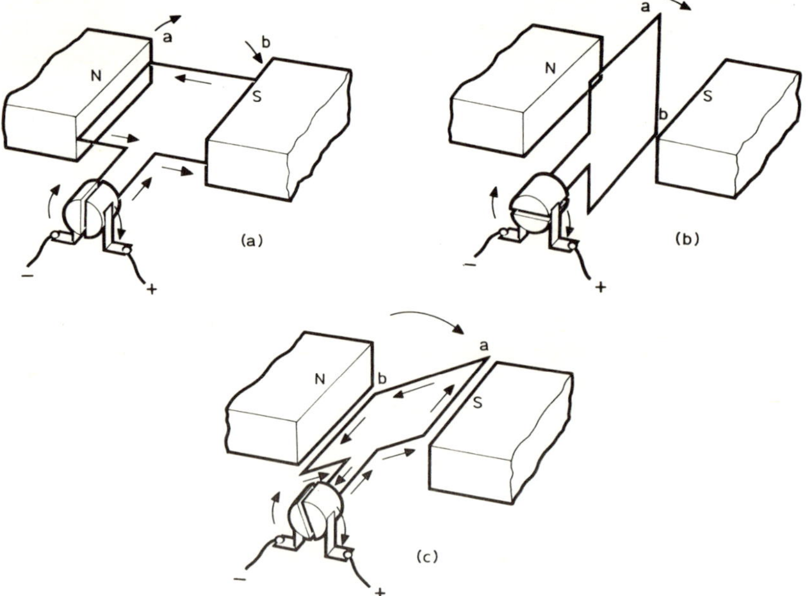

Fig. 7.1 Principle of d.c. electric motor.

These work on the principle shown in the sketch of Fig. 7.1. The current applied to the motor is passed through contacts called brushes to a split cylinder called the commutator, which is fixed to the motor shaft. In the position shown (a) the current flowing in the conventional direction from + to - passes through a brush and commutator segment to flow in one direction round the coil of wire called the armature coil and back out through the other commutator and brush. The combination of the current in the coil and the magnetism of the magnet causes force (more accurately, a **couple** of forces) acting to turn the armature coil round.

In the position (b), the force no longer acts in a direction which will cause turning, because of the position of the wire of the armature coil relative to the magnet. This is not a great worry if the armature has already started moving, because its inertia will keep it turning past this critical position.

In position (c), the new position of the commutator segments causes the current in the armature coil to move in the opposite direction to that in (a), so that the forces acting keep the armature turning in the same direction. Note that the action of the commutator is to keep the coil current moving in the same direction relative to the magnet, though in opposite directions relative to the coil.

Dynamic Action

The arrangement described above like that of any d.c. motor, also acts as a generator; we have in fact used it as a generator in Chapter 4. Because of the dynamo action, another voltage, called the back e.m.f. exists in the circuit when the motor shaft is turning. This back e.m.f. is so named because its direction is opposite to the direction of the voltage which we apply to the motor. As the motor speed increases, the size of the back-e.m.f. increases, and the result is that the current through the motor **decreases** as the shaft speed increases. The reason is that the voltage available to push current through the resistance of the armature coil is the difference between the applied voltage and the back e.m.f.

The speed of the motor becomes steady when the back e.m.f. is nearly equal to the applied voltage; and the difference between the two is used only to pass current through the resistance; this part of the current represents power wasted as heat. When the motor is switched on, the current is much larger than when the motor is running at normal speed (perhaps ten times greater) because there is no back e.m.f. until the shaft rotates.

A motor of this type has a good amount of starting torque, meaning that it can overcome a fairly large load when it starts up because of the large amount of current flowing, assuming full voltage is applied at startup. The motor also tends to run at a fairly constant speed when fed from a steady voltage, because any increase in the load which might slow the motor down causes the back e.m.f. to decrease, so allowing more current to flow and increasing the torque of the motor to cope with the additional load.

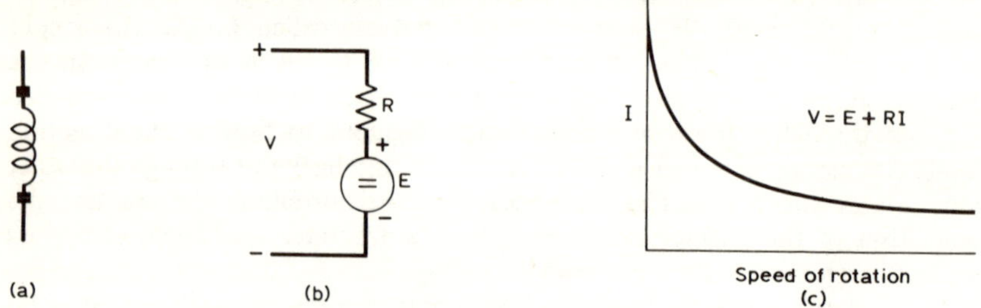

Fig. 7.2 (a) motor symbol, (b) representing the motor as a resistor in series with a generator, (c) motor equation and graph of current against speed.
Example of the practical effect of the motor equation — showing how a load which needs a considerable increase in power (89%) causes only a 10% drop in speed — is worked through below:

Imagine a motor operating from 12V d.c. and passing a current of 100mA at 3 000 r.p.m. Winding resistance = 10Ω

RI = 10 x 0·1 = 1V Therefore, E = 11V at 3 000 r.p.m.

Now imagine speed reduced by 10% to 2 700 r.p.m. by a load. E will decrease in the same ratio to (11 x 2 700)/3 000 = (11 x 9)/10 = 9·9V

Now V = E+RI and filling in the value of 9·9V for E, then 12 = 9·9+RI, so that RI = 2·1V and I = V/R = 2·1/10 = 0·21A (210mA).

The current has more than doubled (110% increase) for a speed reduction of 10%. Original power = 11 x 0·1 = 1·1W. New power = 9·9 x 0·21 = 2·08W (89% increase).

Shunt-wound Motors

These characteristics make the permanent magnet motor ideal for keeping a constant speed and for easy starting, but larger motors using a wound magnet (field) and operated from higher voltage d.c. supplies are used for purposes where the load is variable and the speed must be kept fairly constant. These shunt-wound motors, as they are called, can also be speed-controlled by controlling the amount of current flowing in the field winding.

With less current in the field winding, the motor has to revolve faster to create the same back-e.m.f., so reducing the field current **increases** the motor speed and increasing the field current decreases the motor speed. Shunt wound motors are unlikely to be found in the sizes used for model work.

Speed Control

Unfortunately, most permanent magnet motors used for model work are not intended to run at a steady speed, but to be speed controlled, and there is no field winding to make control easy. The normally-found method of speed control is to use a variable resistor in series with the motor supply, so that the voltage across the motor becomes less as the current increases. Using a resistor in this way makes the current low for starting, so that the starting torque is low, and also makes the speed regulation poor, so that every change in the load needs a change to the resistance value if the motor is to keep running at reasonably constant speed. The result is jerky starting and stopping and poor control of speed, which is particularly objectionable in model rail layouts and in slot-car racing.

When permanent magnet motors are operated from battery supplies, it is not too easy to make speed regulation smooth, but one way is to regulate the voltage across the motor by an electronic regulator. If we apply the voltage which is across the motor to a transistor which is connected as an amplifier, we can use the output of this transistor to control the current through a second transistor which supplies the motor. In this way, irrespective of the load, the motor will (within its limits of power) run at a speed which supplies the correct back e.m.f. to the transistor.

If the back e.m.f. is too low, the transistors act so as to supply more current; if the e.m.f. is too high, the transistors act so as to supply less current, so that the motor is kept at a fixed speed. If we make the voltage supply to the first transistor a controllable variable voltage, then this can be used to control the motor speed. This circuit can therefore be used either to control the speed of a motor or to set a motor to a selected speed when battery or smoothed mains supplies are used.

Practical Circuit

The circuit of Fig. 7.3b should be suitable for most small motors of the types used for model locomotives or slot cars. Tr1 is the current regulator, and the type of transistor used for Tr1 will depend on the amount of power which is taken. For motors operating on 9V and passing up to 50mA, the power

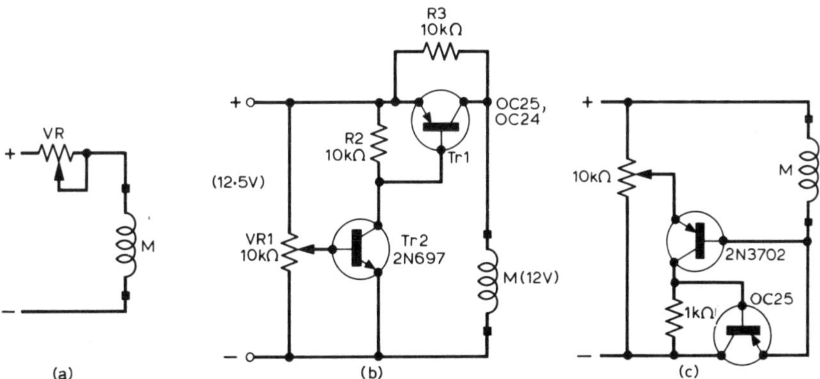

Fig. 7.3 Speed control. (a) series resistor, (b) electronic method, (c) rearranged circuit which makes it possible to bolt the power transistor to an earthed chassis to assist in dissipating heat.

handled by the transistor can be up to 450mW, though it will usually be less than half of this value.

It is better to play safe and use a power transistor which is more than adequate, so as to allow margin for safety if the motor is stalled or, worse still, if the output is shorted. Transistors of the BD131 class are ideal, since they can dissipate several watts and can stand high temperatures caused by large currents flowing.

The voltage across the armature, which is also the back e.m.f., is applied to the emitter of Tr2. When the emitter of Tr2 is 0·5V less than the voltage at the base of Tr2 controlled by the potentiometer VR1, the transistor will conduct, and the voltage at the collector will be low because of the current in the load resistor, R2.

A low voltage, less than the supply voltage, on the base of Tr1, will cause Tr1 to conduct fully, because this is a p-n-p type of transistor which is switched on by a voltage at the base which is less than the voltage applied at its emitter. With Tr1 on, a large current will flow to the motor. When the voltage across the motor, due to a large current off to the back e.m.f., reaches a level about 0·5 less than the voltage of the base of Tr2, any rise in voltage at the emitter will cause Tr2 to stop conducting, so that the voltage at its collector will rise cutting off Tr1 as well.

The action of the circuit is therefore to stabilise the voltage across the motor, making it equal to the voltage set by the potentiometer. Varying the potentiometer VR1 will alter the voltage at which the circuit will stabilise and so alter the speed of the motor. For greater control, the amplification can be increased by using an integrated op-amp in place of Tr2, (Fig. 7.4.)

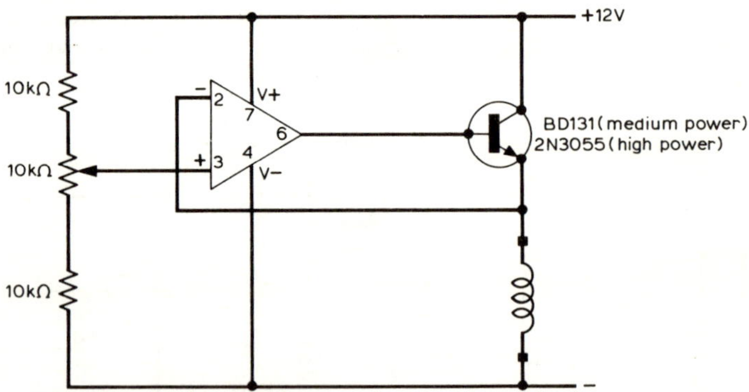

Fig. 7.4 Using an op-amp in the motor control circuit. Note that if the same supply is used for the op-amp as for the motor, the output voltage cannot be reduced to zero.

Stabilisation

The stabilisation of motor speed means that any change in load will cause a change in current so that the torque of the motor is adjusted to the load, within the limits of the motor; a rather higher voltage can be applied to make the most of this, so that a 9V rated motor can be supplied from a 12V battery, though this increases the power dissipation of the transistor.

Because of the improved torque characteristic, the motor will respond more realistically to changes of control setting, and starting and stopping can be made much more smooth. In addition, the control potentiometer can be a much smaller and more easily obtainable type than would be used if the current were being controlled directly. Because the control is through transistors, the motor voltage can be switched on and off, or speed controlled in a mark-space arrangement by logic circuits.

Mains Supplies

When a mains powered controller is used for miniature d.c. motors, the supply is generally not smooth d.c. as would be obtained from a battery, but a waveform called full-wave rectified d.c. The graph of voltage against time for this waveform is shown in Fig. 7.5a. The voltage is always in one direction, unlike a.c., but it varies from zero to maximum and back 100 times per second (with 50Hz mains; the variation will be 120 times per second when the mains frequency is 60Hz). The average value of this wave is 0·64 times its peak voltage, and this average value is the value which will be indicated by a d.c. meter. The average value will also give the same effect on the motor as a d.c. voltage of the same value.

The control circuit of Fig. 7.3 will work with such a supply, and is still useful for the smallest motors, but the use of this type of unsmoothed supply can be an advantage for controlling the speed of larger motors. In the circuit of Fig. 7.3, the transistor Tr1 can handle the full voltage of the power supply and also the full starting current of the motor at the same time, so that it can be asked to dissipate a power equal to supply voltage x starting current.

For small motors, this is no problem, but for larger motors, high-power transistors will have to be used and possibly fitted with heat sinks (see Chapter 8). We can make use of the fact that the unsmoothed full-wave rectified voltage returns to zero 100 times per second, and use a different type of semiconductor called a thyristor.

Thyristor Control

A thyristor looks and acts, up to a point, like a semiconductor diode, passing current in one direction only. Unlike the semiconductor diode, which passes current whenever its anode is more than a few millivolts positive to its cathode, the thyristor passes current only when the anode is positive and when current is also passing to a third electrode called the gate.

When a voltage is applied to the thyristor, anode positive, no current will pass unless the gate has been switched on by a small voltage applied between gate and cathode. Only a small and brief current is needed, and once switched

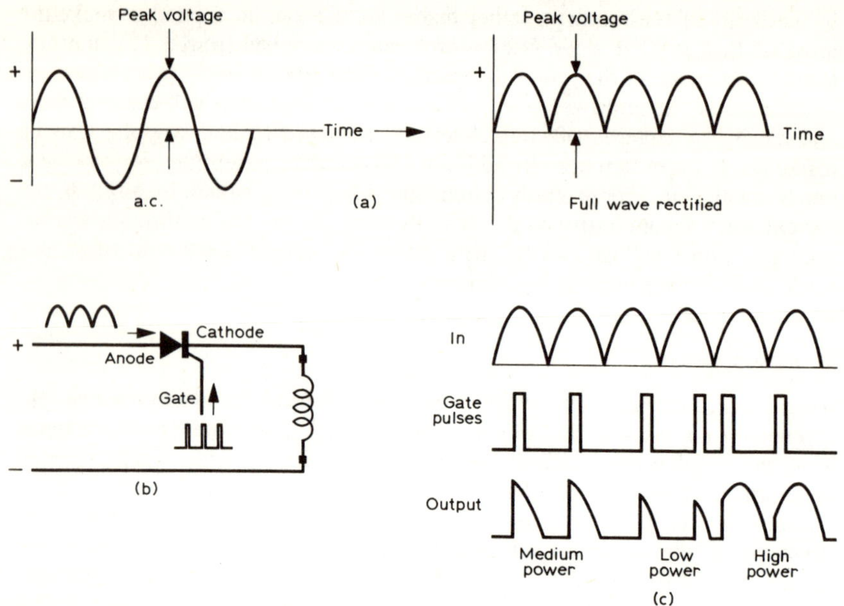

Fig. 7.5 Using thyristors. (a) waveform from rectifier, (b) outline circuit of thyristor control, (c) waveforms.

on the thyristor will remain conducting until either the voltage across it or the current through it has fallen to a very low value, called the minimum holding value.

If we use the thyristor in the circuit of Fig. 7.5b, we can arrange to switch on the thyristor at any part of each pulse of voltage. Provided that the pulse at the gate of the thyristor is brief, or drops to zero at the same time as the anode voltage drops to zero, then the thyristor will switch off at each zero and will not switch on again until the anode is again sufficiently positive and the gate is again pulsed.

We can see the effects of pulsing the gate at different parts in the cycle by examining Fig. 7.5c which shows the effects of early, middle, or late gate switching of each pulse. The effect on the output is to vary the power and average voltage from zero to the normal maximum. The torque can remain fairly high, because of the sharp rise of voltage when the thyristor is switched on, and the supply can be used for any type of d.c. motor, whether permanent magnet or wound field; even for series field motors whose speed is otherwise rather difficult to control. For modelling purposes, the thyristors do not have to withstand high voltages or high currents, so the relatively cheap range of 50V thyristors can be used.

Since thyristors can handle much greater currents and voltages than transistors of comparable size, and with much lower heat losses, because they are either fully on or fully off, they can also be used for control of much more powerful motors, well outside the range used for models but of interest in running power drills and other equipment of interest to modellers. Of particular interest is the application of thyristor control to a lathe so that the chuck speed can be changed smoothly without recourse to gear changing or pulley adjustments. As we shall see later, thyristor type of control can be applied to the types of a.c. motor used for workshop tools.

Trigger Pulses

So far, so good. With a thyristor in series with the motor of the model, we can control speed by triggering the gate of the thyristor with a pulse of the same frequency as the full-wave rectified supply, which can be set to arrive at different times in the cycle. The problem is now to devise a circuit which will provide the trigger pulses. Circuits which operate on the mains directly and are used for mains voltage motors can use very simple trigger circuits, but at the voltages which we are using, these are not very satisfactory.

A better method is to use a type 555 timer circuit to generate a time delay. The 555 is triggered by the supply waveform, and generates a squarewave whose duration is controlled by the potentiometer VR1. The steep end of the output squarewave is picked off and used to fire the thyristor. This trailing edge of the squarewave will occur some time after the start of triggering, so that altering the delay will cause the thyristor to fire at different times in the cycle. Fig. 7.6 shows the complete circuit.

Though thyristor control is effective for altering the speed of motors, its regulating action is not so good as that of the transistor circuit which was discussed at the beginning of this chapter, because there is no way of keeping the

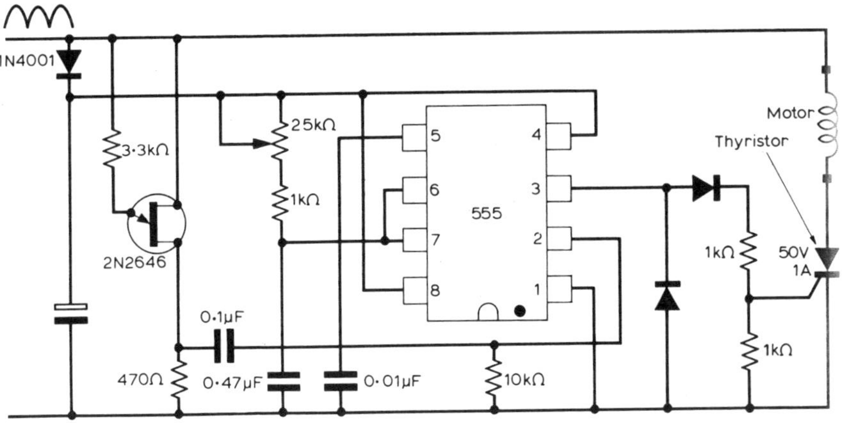

Fig. 7.6 Control circuit using a thyristor. If the power supply unit can be modified, the circuit of Fig. 7.7 is simpler and more effective.

voltage, even the average voltage, of the motor constant at a constant setting of the regulator. If better regulation is needed, the voltage across the motor must be used to control the firing of the thyristor.

This is done in the circuit of Fig. 7.7 by wiring the motor between the cathode of the thyristor and earth, so that a voltage greater than the motor voltage must be applied to the gate to fire the thyristor. If the motor voltage exceeds the voltage across C1, the thyristor will not fire; if the voltage across the motor is low, the thyristor will fire early in the cycle. The voltage to which C1 charges during half of the cycle is controlled by VR1, so that this acts as the speed controller.

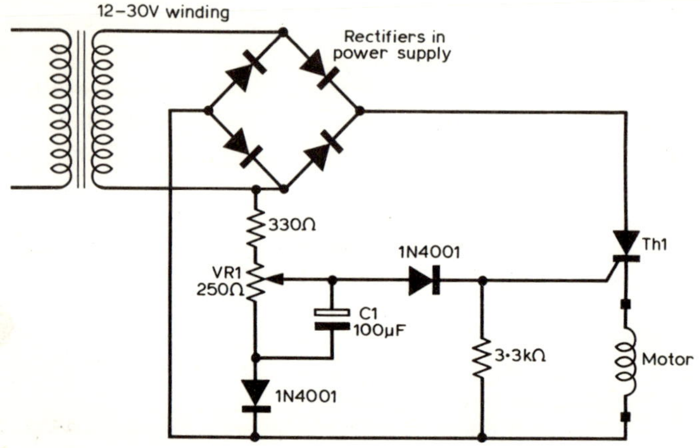

Fig. 7.7 Simpler control circuit which can be used if the additional connections can be made inside the power supply. The thyristor Th1 can be an RS type BTY30-25 or any thyristor rated at 50V and full rated (peak) current (1·4 x maximum motor current).

Controlling Mains Motors

Workshop tools, particularly electric drills, operating from a.c. mains can also be controlled by using thyristors and the more modern triacs. The triac is a two-way thyristor which is capable of controlling current in both directions so that it is more suitable for use in a.c. control systems. The circuits used for the control of electric drill motors up to 1kW power output are comparatively simple, but because they work at mains voltage more care is needed over their construction. In particular, younger modellers should not attempt these circuits nor those of the mains powered voltage supplies described in the next chapter.

A simple drill speed controller is shown in Fig. 7.8. This controls only half of the range of power of the drill, so that a switch has to be fitted to add the remainder of the range. A single thyristor is used, and its gate is triggered by the mains waveform attenuated by the components R1, VR2 and the variable

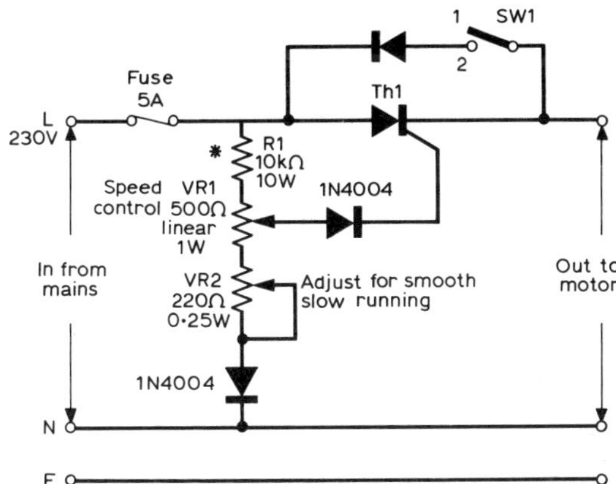

Fig. 7.8 Speed controller for electric drill (500W approx.). For higher power drill motors, a thyristor rated at 10A should be used. In this circuit Th1 can be a type BTY79-400R or BT109 for 500W load, using heatsink rated at 4°C/W. Rating of these thyristors is 400/500V 6A. VR2 is adjusted so that the motor speed is steady at the lowest setting of VR1. Sw1 positions are 1 = 0-1/2 speed; 2 = 1/2-full speed. (*) Note that resistor R1 will run hot at all times when the unit is switched on and care must be taken to mount R1 clear of all other components, wire or plastics materials. Some ventilation is needed, as the thyristor will also run hot, and a heatsink is needed. (Circuit by courtesy of RS Components Ltd.)

resistor VR1. This simple method can be used because the voltage needed to trigger the gate is only a small fraction of the mains voltage across the network R1, VR2 and VR1. At lower supply voltages, the triggering voltage would be achieved only when the supply voltage was near its peak. Switch Sw1 is used to set the ranges 0 to 1/2 power and 1/2 to full power.

Insulated Construction

Two forms of safe construction may be used. One is to insulate the circuit completely, mounting the components in a (ventilated) plastics case with only the plastics shaft of the potentiometer VR1 and the plastics knob of the switch taken through the front panel. The other method is possible only when the controller will be used with three-core cable and plugged into three-pin sockets which include a good earth connection. If the controller is to be mounted on the drill cable, remember that several types of drill use double insulation, so that there is no earth lead in the cable. In this case, a completely insulated construction must be used for the controller also.

A good method for insulated construction is to use two plastics boxes which fit one inside the other; suitable boxes are used for storage of deep frozen food. The controller circuit is built inside one box, securing the potentiometer switch and the rest of the circuit to this box, and then mounting the whole assembly into the second box, using non-metallic fasteners.

The shaft of the potentiometer and the knob of the switch will protrude through holes cut for them in the second box, but there is no accessible metal part which can be touched. This requires the use of a potentiometer with an insulated shaft and a switch with an insulated knob. Make sure, during construction, that stranded cables used for mains connections are soldered at the tips before they are fastened in. This lessens the chance of a strand of wire straying out of place and putting mains voltage where it is unwelcome.

Metal Box Construction

The controller can also be made up in a metal box, such as the *Veroboxes*, or the range available from RS supplies or from Norman Rose. The box should be securely earthed through a three-core mains cable, and should carry a three-pin socket so that drills or lamps of up to 1kW total power can be plugged in. With this type of construction, it does not matter whether the lead to the drill is two-core or three-core provided that the controller itself is properly earthed, with its plug carrying a 7A fuse and plugged into a properly-earthed three-pin socket.

In the construction of the controller, metal potentiometer shafts and switch knobs may be used only if the controller is built into an earthed metal box. Take great care, with any form of construction, to avoid small gaps between line leads and other points in the circuit, since shock or vibration could cause components to shift enough to bridge a small gap. Use sheets of plastics material, well secured to prevent any possibility of the circuit board touching against metal casing, particularly if it were to break loose from its leads.

The construction should be aimed to withstand typical hazards, which might include dropping from a 3-ft high work surface on to a concrete floor, or being run over by the car. In either case, nothing should happen which can make the accessible parts of the controller live to mains.

CHAPTER EIGHT

POWER SUPPLIES

POWER SUPPLIES ARE NEEDED both for electrically operated models and also for electronic systems; and these power supplies may be battery or mains operated. In general, the amounts of current taken by model motors, solenoids and other equipment are large enough to make the use of batteries, other than the rechargable types, a rather expensive proposition. In some instances, however, such as radio control receivers and transmitters, there is no option to the use of batteries.

Electronics circuits can be made to consume very small amounts of current, so that battery operation of some types of circuit is perfectly satisfactory. Circuits using TTL logic or counting elements are best operated from mains supplies because of the comparatively high current which they take, and circuits which operate solenoids can be operated from the same mains supply which will be needed for the solenoids.

An important point to remember is that low voltage battery supplies are completely safe from the point of view of electric shock; correctly designed and constructed mains supplies are also very safe, but poorly designed, constructed or operated mains supplies can be a death-trap. For this reason, we feel that younger modellers should use only commercially constructed mains supplies or keep to battery power for all their modelling activities involving the use of electricity. Older constructors who tackle the mains power supplies outlined in this chapter should heed carefully the advice on construction and layout, particularly with regard to insulation and earthing.

Dry Cells

Batteries today are available in a bewildering assortment of shapes, sizes and types, but we can divide them into rechargable and disposable. Disposable batteries are (like any other batteries) assembled from cells, with each cell giving a voltage which is about 1·5V for the most common type of dry cell. The most common cell is the dry Leclanché (zinc-ammonium chloride) as fitted to countless bicycle lamps and transistor radios. This cell is comparatively cheap, has a reasonable shelf life, and can be obtained almost anywhere. Its disadvantages include a rather high internal resistance, short life when large

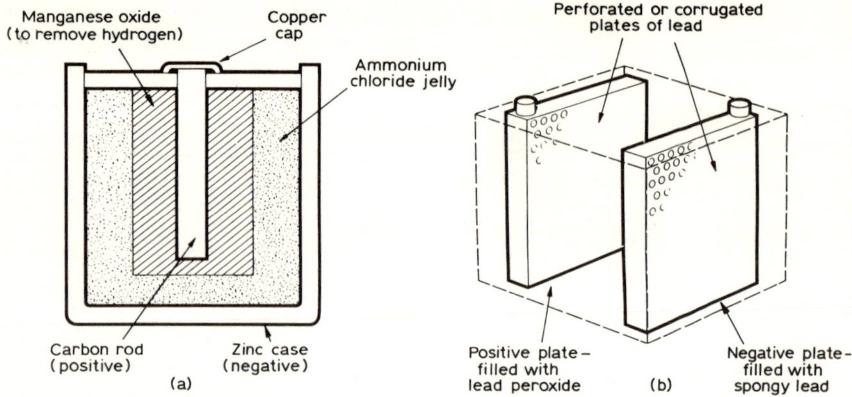

Fig. 8.1 (a) Leclanché dry cell section, (b) lead-acid accumulator cell.

currents are taken, and the risk of leakage when the cell is exhausted or has been left for some time unused.

High internal resistance means that the voltage across the terminals of the cell, nominally 1·5V for a fresh cell, will drop considerably when current is taken from the cell, reaching a voltage of 1·2V or less when the current is 500mA. Small cells will have higher internal resistances than large cells, so that currents of 100mA or so will cause a noticeable drop in voltage. High currents taken for any length of time also cause rapid exhaustion of the cell, so that the equipment powered by the cell stops working.

After some recovery time with the current switched off, the cell will recover its original voltage, but will probably have a greater value of internal resistance. Once again, physically small cells suffer more from this fault than the larger ones. When the cell is at the end of its life, the zinc casing which forms the outer casing of the cell is very thin, and there is a danger of the corrosive jelly inside the cell leaking out, causing damage to any metalwork round the cell. A common fault in battery-operated equipment is severe corrosion of the battery contacts caused by leakage from the cell.

Battery of Cells

The simple dry cell, which is also available as a "leakproof" cell with a steel case round the zinc, has a voltage, on no load, of 1·5V. Making a battery of such cells in parallel causes the total internal resistance to drop, with the total voltage still 1·5V, so that more current can be drawn at this voltage. If cells are connected in series, the total voltage is the sum of all the cell voltages, which is 1·5 x number of cells. In this way, the common battery voltages of 3V, 4·5V, 6V, 9V, 12V, 15V and so on, are obtained.

In any equipment operated from dry batteries, the use of a physically larger battery is always an advantage, provided that it can be fitted in. For example, a pocket transistor radio operated from a small PP3 type of battery may have a battery life of a month or so; if the larger PP9 type of battery is used the life can be as long as a year, so that the greater cost of the larger battery is repaid several times over, but the battery must be strapped to the outside of the radio since it is too large to fit inside.

Other Disposable Batteries

Mercury cells and batteries, and the manganese-alkaline types can be bought from most chemists which supply photographic goods. These types of batteries are used where the higher price is offset by their special advantages. Mercury batteries can be made in very small sizes with high current capabilities and comparatively long life; they are used in deaf-aids, in the automatic exposure controls of cameras and in digital watches. Manganese-alkaline batteries can provide large currents for short times and have a long storage life when not in use; they are used in equipment such as flashguns. Though these batteries are very useful in radio control equipment, we do not make very much use of them for general modelling work.

Lead Acid Accumulators

The most common type of rechargeable cell is the lead-acid accumulator used in car batteries and in its smallest forms for powering model aero glow-plugs. As the name suggests, each cell consists of lead plates dipping into sulphuric acid. One plate, the positive one, is coated with lead peroxide, the other with spongy lead, and when fully charged the voltage of each cell is 2·2V, so that a 6-cell battery, nominally 12V, gives 13·2V when it is fully charged, and the voltage is greater if the cell is connected to a generator and is still on charge.

As the cell discharges, the lead peroxide reacts with the sulphuric acid to form lead sulphate, and the voltage drops very quickly to about 2V per cell, and then remains almost constant until the cell is almost discharged. At the same time, the density of the acid decreases, so that the state of each cell can be checked by measuring the density of the acid using a hydrometer, assuming that the acid was of the correct strength to start with.

The density of a new battery should be checked whenever the cell is fully charged, as low-strength acid will make the life of the cell rather short. The only remedy is to return the cell, for sulphuric acid cannot be purchased easily, and is undesirable to have around the garage in any case.

A discharged battery can be recharged by passing current into it from a supply of d.c. at higher voltage, connected with the + pole of the supply to the + terminal of the cell. Such a mains charger can be simply built (see later). If a lead acid cell is left discharged for long periods (several weeks) and particularly if it is left at low temperatures, the lead becomes coated with lead sulphate on both plates, recharging becomes impossible, and the cell must be renewed.

The lead-acid accumulator has a high voltage per cell, and can pass large currents because its internal resistance is low. It is heavy, and the acid which it contains is dangerous. Smaller cells, such as those sold for glow-plug starting, contain the acid in a jelly form so that spillage is less likely, making the cell much more convenient to carry around and use.

Nickel-Cadmium Cells

Nickel-cadmium cells are the most recently available type of rechargable cell. Their voltage per cell is fairly low, about 1·2V charged, and they are expensive, but they can be made in fairly small sizes and have fairly long lives. Since the cells are sealed, recharging must be carried out carefully, following the manufacturer's instructions, since the pressure of gases released during charging could break open the case.

Mains Supplies

For fixed supplies, or for charging accumulators of the lead-acid variety, some form of mains supply is needed. Mains power supplies can consist of three sections; a transformer, a rectifier, and smoothing. The transformer changes the mains voltage of 230V a.c. into a lower and incidentally safer value. The rectifier converts this low voltage a.c. into current flowing in one direction, though by no means pure d.c., and the smoothing converts this in turn into pure d.c. For many purposes, such as battery charging, and also to some extent for motors, no smoothing is needed and the simple circuit of Fig. 8.3 is used.

Because of the way in which a.c. quantities are measured and the effect of taking an average value of a voltage which is not steady, the relation between the transformer output voltage and the d.c. output voltage is not exactly obvious. For the circuit drawn, the d.c. output with no current flowing is 0·9 x the voltage (r.m.s.) of the transformer.

R.M.S. means "root-mean-square", the method used for measuring the "average" value of a.c. This does not tell the whole story, because the peak value of an alternating voltage will be equal to 1·4 times the r.m.s. value. For example, a 15V r.m.s. transformer (we shall cease to use r.m.s., since all the quoted voltages of a.c. are r.m.s.) is very often used to charge 12V accumulators, since 15 x 0·9 = 13·5, which is very close to the 13·2V of a fully charged

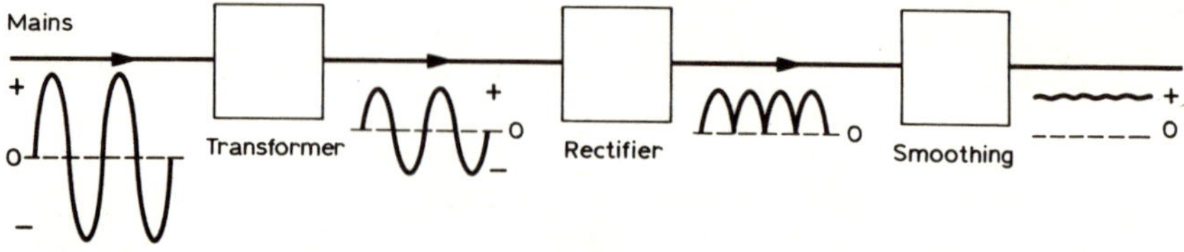

Fig. 8.2 Block diagram of mains power supply.

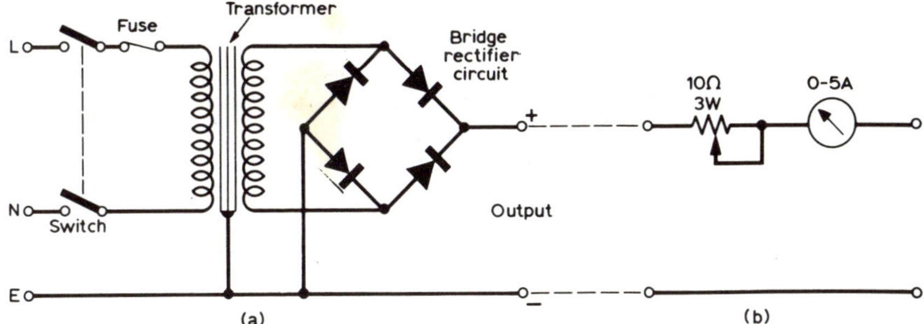

Fig. 8.3 (a) simple battery charging circuit, (b) variable resistor and meter added for 'de-luxe' version.

battery. The peaks of voltage are 15 x 1·4 = 21V however, so that some current will continue to flow into the battery even when it is fully charged.

The simple battery charger relies on the resistance of the transformer windings to avoid passing excessive currents during these peaks; slightly more elaborate chargers use a variable resistor on the a.c. side to control the maximum current flowing into a discharged battery. The most elaborate types of charger circuits incorporate electronic controls to limit current and to cut down charging current to a trickle when the battery is fully charged. This has the advantage of reducing the loss of gases from the accumulator.

These gases come from the splitting of water into hydrogen and oxygen, and mean that water is being lost from the cell. This water must be replaced in the form of pure distilled or otherwise demineralised water to keep the level of liquid in the cells just at the top of the lead plates. Because the mixture of

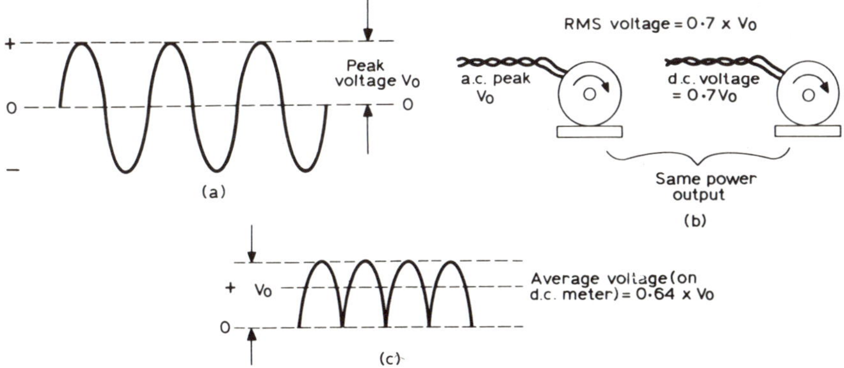

Fig. 8.4 A.c., d.c. and rectified voltages. (a) a.c., showing peak value, (b) r.m.s. and peak voltage which would run an a.c. – d.c. motor at the same power output as the a.c. The r.m.s. voltage = 0·7 x peak voltage. (c) rectified a.c. – the average value measured on a d.c. meter is 0·64 x peak voltage.

oxygen and hydrogen is extremely explosive, great care should be taken when cells are being charged.

Never smoke near a charging cell, and never disturb the connections when current is passing, either charging or discharging, in case of a spark. Ignition of the gases causes a violent explosion which bursts the battery case and showers everything around with sulphuric acid. The main hazard is to the eyes of the operator. In such an emergency, copious amounts of water are the first aid treatment; but expert advice may be needed if sight is to be saved.

Power Supply Precautions

In the basic battery-charger design, as in all mains power supplies, great care must be taken over electrical safety. If, as is usual, the supply is built into a metal case, the earth (green/yellow) wire of the supply cable must be used to connect the casing to earth. The earth wire may be soldered directly to the case if the case is steel, or soldered to a tag which is then bolted and lock-nutted to the case. The mains cable must pass through a rubber grommet on its way into the case so that it cannot be cut by rubbing against the edges of the hole, and must be clamped inside the case so that pulling on the cable cannot detach the electrical connections.

The live (brown) and neutral (blue) leads should be taken to the mains switch, then the live lead from the mains switch to the fuse, and from the fuse to the transformer primary. The other primary connection is then taken to the other terminal of the switch, assuming a two-pole switch. All of these leads should be run in well-insulated wire, with as little of the insulation stripped back as is essential for soldering connections.

The cover of the case should be constructed so that no live point can be reached through any ventilation gap, even with a very thin finger. The live and neutral leads should be shorter than the earth lead, so that if the worst happens the live leads will break before the earth does. At the plug end, the wires must be correctly attached and the plug fitted into a correctly wired socket. Provided that this 230V wiring is correctly carried out, kept well away from other wiring or metalwork and out of reach of fingers, electrical safety should be good.

If there is any likelihood of servicing or adjustment being done with the case off, all exposed 230V points must be covered, neutral as well as live. Insulating tape should be regarded for what it is, a temporary exedient only; the best protection is a blob of silicone rubber (sold for sealing the rims of baths) round the live terminals of the switch, fuse and transformer, and a small plastics box lid glued over the transformer terminals. The open type of fuse-holder should never be used at mains voltages. The low voltage part of the supply can then be constructed like any other electronic circuit.

Note that if a power supply is built in a plastics or other insulating box, the earth wire from the mains lead should be secured to the metal frame of the transformer, and the negative output lead also taken to a tag fastened to the transformer case. This ensures safety if a transformer winding fault should develop. Never omit earthing of a power supply simply because the supply is built into an insulated case.

Supplies for Electronic Circuits

The simple supplies using transformers and rectifiers alone are unsuitable for most electronic circuits except these using thyristor controls. For most of the electronic circuits described in this book, smoothed supplies are needed, with no remaining a.c. ripple. The simplest method of providing smoothing is to add a capacitor across + and - supply leads, and this also causes the voltage to rise to 1·4 x transformer output volts when no current is taken, because the capacitor stores the peak value of voltage.

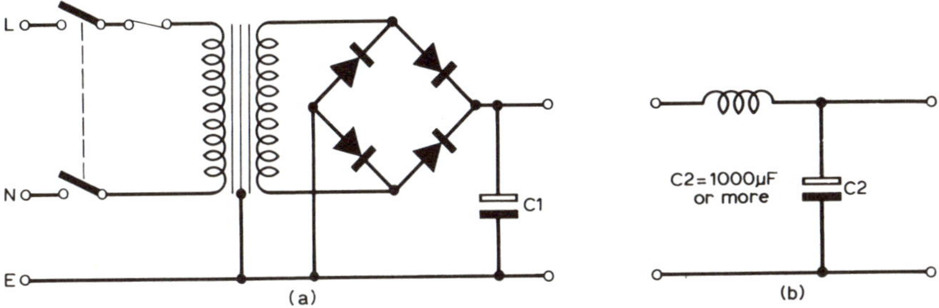

Fig. 8.5 Smoothed output power supply: (a) basic circuit, (b) added filter for very smooth d.c. The ratings for components are important to observe, as follows:
Rectifier: the rating should be for reverse voltage equal to at least 1·4 x a.c. output of transformer. (for 15V transformer, V_{RRM} = 1·4 x 15 = 21V). Current rating = 1/2 x output current (for 1A output, each rectifier passes 1/2A average).
Capacitor C1: This should be 1,000µF or more. Voltage rating should be 1·4 x a.c. output of transformer at least. For example above, use 25V rating minimum.

The voltage will drop and the ripple will increase as more current is drawn, but it is not simple to predict what the measured values of voltage will be. For parts of circuits which need a really well-smoothed supply, an additional filter can be used which will cause a further voltage drop. If the voltage must be held at a constant value, then stabilised circuits must be used.

TTL Circuits

Electronic logic and counting circuits using TTL integrated circuits must have a supply of 5V nominal, which must not rise above 5·25V if normal action is to be guaranteed, and which must not rise above 6V or so if damage is to be avoided. A 6V battery, whether dry or accumulator, gives too high a voltage, but an accumulator can be used with a voltage reducing circuit of the type shown is Fig. 8.6 if only a few TTL circuits are to be operated.

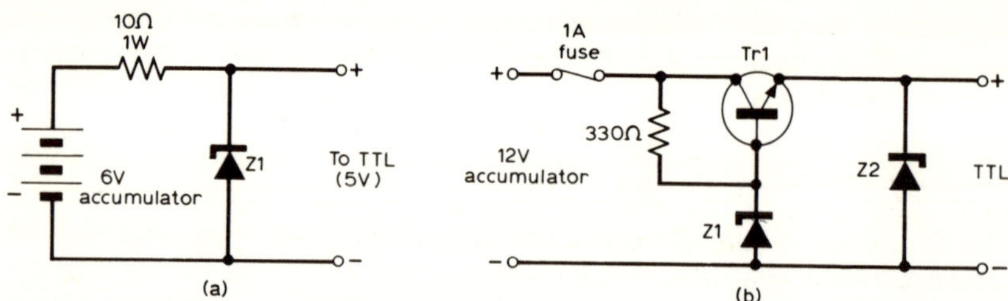

Fig. 8.6 TTL supplies using accumulators. (a) simple circuit for 6V accumulator supply for up to three TTL units. Z1 = 5·1V zener diode, 1·3W dissipation rating. (b) circuit using transistor stabiliser and 12V accumulator. Z1 = 5·6V zener (RS type BZY58-4·3); Z2 = 5·6V zener diode, 1W rating minimum; Tr1 = power transistor to suit current taken (BD131 is suitable, without heatsink, for currents up to 400mA). The zener diode Z2 is used as a safety precaution to prevent the voltage across the TTL cirucits rising to 12V if the stabiliser fails. The fuse would blow before the zener diode overheated, provided that a zener diode of adequate rating is used.

If the power dissipation can be tolerated, a 12V accumulator gives a more reliable 5V supply. A 5·6V zener diode should always be wired across the power supply of a TTL circuit as an additional safeguard in case of failure of other voltage stabilisers. Mains supplies present more of a problem because of the surges of voltage which occur when switching on and off, and a stabilised supply is essential if TTL circuits are to be operated from a mains supply. For a small number of circuits, a simpler zener diode and transistor stabiliser can be used, as shown in Fig. 8.6.

Positive-Negative Supplies

Some electronic circuits using integrated operational amplifiers such as the 741 need both positive and negative supplies balanced around earth. This can be

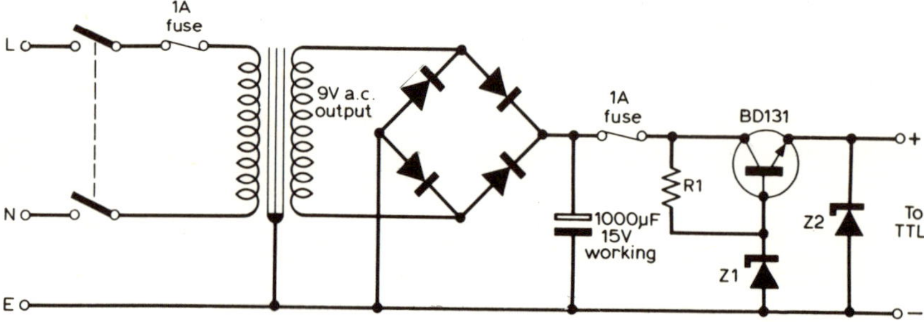

Fig. 8.7 TTL Supply using a.c. mains. Z1 = 5·6V zener diode, 0·5W or less; Z2 = 5·6V zener diode, 1W or more (protective only); R1 = 390Ω, 1/2W. The BD131 transistor should be attached to the metal case of the power supply using the mica washer provided and with a smear of heatsink grease on each side of the mica. The BD131 must not be mounted directly to earthed metalwork.

supplied by two batteries, as shown in several of the circuits in this book, but can also be provided from a mains supply, using the circuit of Fig. 8.7.

Note that both supply lines need to be smoothed by connecting a capacitor to the earth line, and the polarity of these capacitors must be carefully observed. Particular care must be taken with some types of electrolytic capacitor which have a case connected to the negative terminal. When such a capacitor is used for the smoothing of the negative supply, the case must be insulated from the metalwork of the power supply chassis.

Regulated Supplies

Current regulators are frequently very useful in any equipment in which there is a danger of drawing excessive current and damaging the power supply or other electronic components. Fuses, although essential in mains leads, are of little use in preventing damage to transistors or diodes, since the semiconductors will burn out long before the fuse has had time to melt. For this reason, many electronic power supplies are fitted with current limiters or contact breakers.

Current limiters automatically keep the current flowing to a predetermined safe value even if the output is a short circuit. Circuit breakers go rather further and disconnect the circuit, showing a warning light, until a "reconnect" button is pressed. The current limiter has the advantage that the supply is kept on and will return to normal whenever the short is removed, and that no damage can be caused in the way that can happen with a normal contact breaker if the "reconnect" button is pressed before the short is removed.

Current Limiter Circuit

The circuit of a typical current limiter is shown in Fig. 8.9. Tr1 is a power transistor capable of passing the full current which is to be the maximum

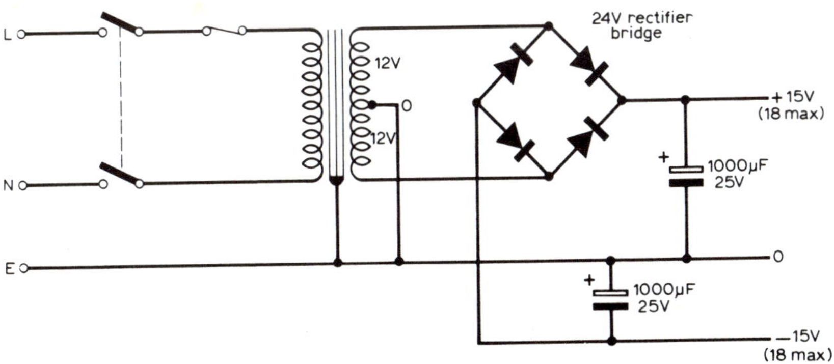

Fig. 8.8 Balanced power supply using a 12-0-12V transformer (24V centre tapped).

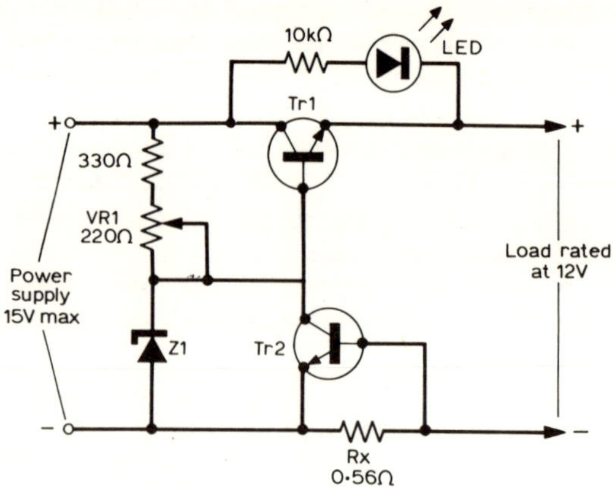

Fig. 8.9 Overload protection circuit. The values shown are for 12V with a maximum of 1A short circuit current. Tr1 can be a 2N3055 or similar power transistor. Tr2 can be a 2N697 or similar general purpose transistor.

supplied, and also capable of dissipating the power equal to supply voltage x maximum current when a short circuit occurs. If, for example, the supply voltage is 12V, and a maximum of 1A is to be supplied, the transistor must be able to dissipate 12W when the output is shorted.

In normal use, the voltage across the transistor Tr1 is very small, not enough to light the LED wired across Tr1. The small resistor Rx is chosen so that the selected maximum current flowing through Rx will cause a voltage of 0·5V. The formula to use is Rx = 0·5 x Imax, where Imax is the maximum current in amps. For 1A, the value of Rx is 0·5 ohm, and the power rating of the resistor should be 1W.

When 1A flows, and 0·5V is dropped across Rx, Tr2 starts to pass current with the effect that the voltage at the base of Tr1 drops, so cutting down the current in Tr1. VR1 provides some adjustment of the exact cut-off current. When the cut-off acts to limit the current, the voltage across Tr1 will rise, and the LED will light to show that excessive current is flowing. Removing the fault will extinguish the LED and restore normal operation.

Voltage Regulated Supplies

Voltage regulated supplies are seldom needed for modelling use, but are useful if a variety of power supplies have to be used for electronic equipment. These need not be built, since they can be bought as integrated circuits, at least in the smaller sizes. The typical circuitry needed with one of these regulators is shown in Fig. 8.9.

For a 5V output for TTL circuits, the RS type 7805 regulator will take an input voltage of 7V to 25V, so that 12V accumulators can be used. The maximum short-circuit current is 750mA, more than adequate for any of the TTL circuits used in this book. The 12V regulator, type 7812 takes 14·5V to 30V inputs, and has a short-circuit current of 350mA. These regulators are mounted in a case which can be bolted to a metal chassis to dissipate heat.

Heat Sinks

Wherever power transistors or similarly-constructed semiconductors (such as some thyristors and triacs) are used to dissipate more power than they can handle with normal cooling, we may have to make use of heat sinks to transfer the heat which has to be dissipated. These heat sinks can be small metal tabs which clip on to small transistors for low powers (about 0·5W) ranging up to large aluminium fins on which power transistors dissipating 25W or so can be mounted.

The important feature of a heat sink is its thermal resistance in units of °C/W. For example, a heatsink of thermal resistance 4°C/W would have a temperature rise of 4°C for each watt of power dissipated. If we used this on a transistor which was dissipating 12W, the temperature rise of the heat sink would be 48°C above room temperature, so that the temperature of the case of the transistor would be (room temperature + 48°C).

If we take the air temperature as being anything up to 35°C (inside the casing where the transistors may be mounted), then the temperature of the outside of the transistor case could be 83°C. This is not the whole story, because the power is dissipated inside the transistor, and has to flow out of the casing, so that the inside is hotter still.

Manufacturers provide figures for the average thermal resistance of their power transistors; for example the thermal resistance of the BD132 is 6°C/W, plus another 1°C/W when it is mounted on to a metal base, or another 4°C/W if it is mounted on a mica insulator. With an insulator, this could add another 10°C/W, so that the temperature inside the transistor with 12W dissipation would be 12x 10 = 120°C higher than the casing temperature which we have calculated as 83°C. This gives a total of 203°C, which is far too great for this transistor, which can handle only 125°C inside.

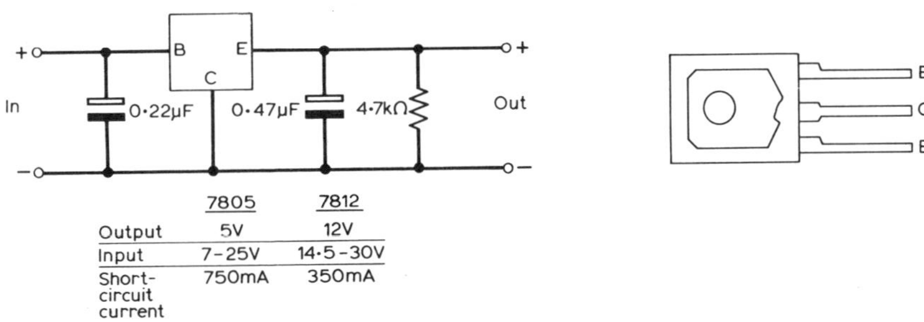

	7805	7812
Output	5V	12V
Input	7-25V	14·5-30V
Short-circuit current	750mA	350mA

Fig. 8.10 Using an integrated circuit regulator such as the RS type 7805 or 7812.

The thermal resistance of transistors and heat sinks will therefore decide how much power can be dissipated safely without destroying the transistor. Since the heat sink requirement will depend so much on the amount of power which has to be dissipated, no definite heatsink sizes can be given for the uses of power transistors in this book, but the circuits where heatsinks may be needed are indicated.

Note that in the power regulation circuits where transistors are used we do not apply full voltage and full current at the same time, so that the power rating of the transistor, apart from short circuits, is only 1/2 voltage x 1/2 current.

Insulation

Most power transistors have the metal casing connected to the collector, so that attaching a transistor to a heat sink automatically connects the heat sink into the collector circuit unless an electrical insulator is used. Kits of mica washers can be bought for all the common power transistors, and have a thermal resistance of about 4°C/W.

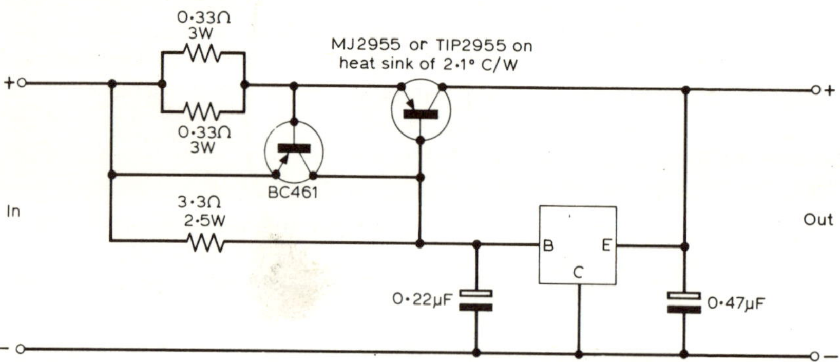

Fig. 8.11 Using this circuit, the regulator can be used for current of up to 3A. (Circuit by courtesy of RS Components Ltd).

If the cooling is more important than the insulation, for example where the heat sink itself can be insulated from other metal parts of the circuit, then it is better to bolt the transistor directly to the metal heat sink, using silicone grease to ensure good thermal contact between the surfaces. Silicone grease should be used also if mica washers are used.

In a few designs, it is possible to amend the circuits so that the collector of a transistor which dissipates most power can be at earth voltage and so the transistor can be mounted directly on the metal work of the case.

CHAPTER NINE

PRINTED CIRCUIT BOARDS

CIRCUITS WHICH USE ONLY ONE transistor and a few other components, and which are made as "one-offs" can most simply be built on *Veroboard*. Even quite complex circuits can be built in this way, and special types of *Veroboard* can be obtained for mounting integrated circuits.

When large numbers of circuits are produced for commercial purposes, the board is specially made for the circuit, starting with an insulated board completely covered with a sheet of copper. The practical layout of the circuit is then copied on to the copper photographically, using a light-sensitive varnish, and the unwanted copper is etched away by an acid solution. As a result, the board has copper strips designed to suit the components, and more compact and reproducible layouts can be achieved.

Lacquer Resist

The photographic method is not suitable for a very small number of boards, but when a few copies of a circuit board are needed, perhaps for club purposes, then the use of etched boards is attractive; and hand or stencil application of an acid-resistant lacquer is suitable. In addition, copper etching is an excellent method of making decorative nameplates.

Most lacquers are etch-resistant, but a coloured lacquer which is soluble in meths or carbon tetrachloride (**Warning** – *Vapour is heavy and has anaesthetic effects*) is preferable, since it can then be removed easily after etching. The copper should be cleaned thoroughly, using a wet paper tissue and domestic abrasive such as *Vim*, then rinsed well and dried. Complete drying is important, as any trace of water under the lacquer will cause the etch solution to undercut the lacquer.

The layout is then drawn out on the copper with the lacquer, using a fine brush or an old felt-tip pen dipped in lacquer. For a large number of identical boards, a stencil of the PCB pattern cut out of plastic sheet is most useful. The lacquer layer should be thick and allowed to dry completely.

Etching

Etching is carried out with Ferric Chloride (Iron (III) Chloride) solution. Dry Ferric Chloride is readily obtainable (see advertisements in *Practical Wireless*,

Practical Electronics, Electronics Today International) and should be made up by dissolving 500 grams of the solid in 1 litre of water, and stirring until the solution is clear, with no trace of crystals. Use plastics containers and stirrers. Solution is much faster if warm (*not* boiling) water is used. Five minutes of dipping the board, held in plastics tweezers from a photographic dealers, in this solution should result in a good etch, though the time needed will be much longer if the solution is cold.

When the board is satisfactorily etched, the acid is washed off with warm water, and the board dried. The remaining lacquer is now removed with the solvent, and the board again washed thoroughly and dried. The holes for the component mounting wires can now be drilled and cleaned up. Finally, the components can be inserted and soldered in place. Remember that the components are placed on the opposite side of the board from the copper.

If any difficulty is experienced with lacquers, it is possible to buy lacquers specifically made up for etch-resist work, and R-S Components, along with many other electronics suppliers, sell a fibre-tip pen ready loaded with acid-resist lacquer. Plain copper board and ferric chloride powder are also readily obtainable from electronics suppliers.

PCB Layouts

The following pages illustrate some typical PCB layouts. Two types of layouts are possible, one using components flat on to the board, the other using upright components. The first method results in a circuit which is thin but with greater area than the second. Whichever is used must depend on the space into which the PCB must fit. Components which may run hot should always be mounted on long leads to avoid overheating the board, or alternatively mounted away from the board altogether.

In general, heavy components such as mains transformers should not be mounted on PCBs, though the PCBs may be mounted on the mains transformer, and the mains transformer bolted to the case. A useful provision in a PCB is an earthed strip all round, with four mounting holes, so that the PCB can be secured to a case and simultaneously earthed. Another useful point is to etch in connection points for external leads along with reference letters or numbers, so that connections to the board can be quickly identified.

Finally, never cut a stencil and start quantity production of a PCB design until the first-off board has been used for construction of the circuit, mounted in place and thoroughly tested. One failure is tolerable, 20 makes life tedious. Good etching!

PCB Patterns

The patterns shown in the drawings which follow are drawn *twice full size*. In each case, flat-on-board mounting has been assumed, along with components of standard dimensions. Inevitably, if a mixed bag of components is used, some will not readily fit the board, but most can be accommodated with a little wire-bending. The patterns are drawn as seen from the copper side of the board.

Resistors – 1/4W carbon or film.

Capacitors – rated at just over maximum working voltage of the circuit

Transistors – all except power types use 2·5mm (0·1″) wire spacing.

ICs – Use 2·5mm pin spacing

Preset variables – use open types with pins on a 5·1mm pitch circle radius. Transformers and solenoids and other large components are not mounted on the boards.

Not all the circuits in this book are shown in pattern form. These selected are circuits which might have to be made up in fairly large numbers, or which need a PCB for a neat layout. Particular care must be taken with the PCB patterns for ICs, since fine connecting lines are needed, and there is a risk of shorting from one line to the next.

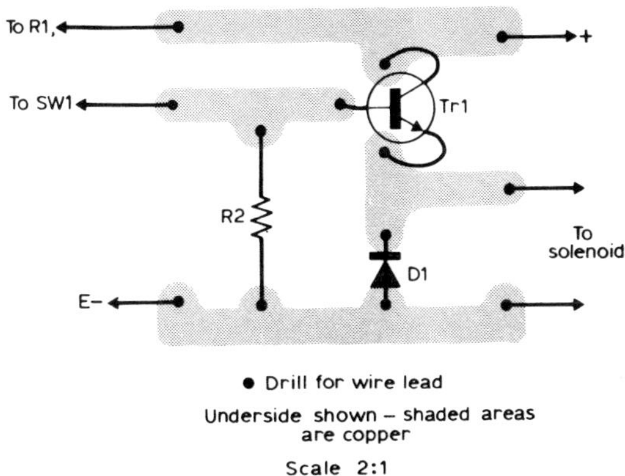

Fig. 9.1 PCB layout for solenoid/relay control circuit of Fig. 2.1 This layout, and all others shown in this chapter, is reproduced twice full size and is shown from the copper profile side with components viewed as they would be seen through the laminate.

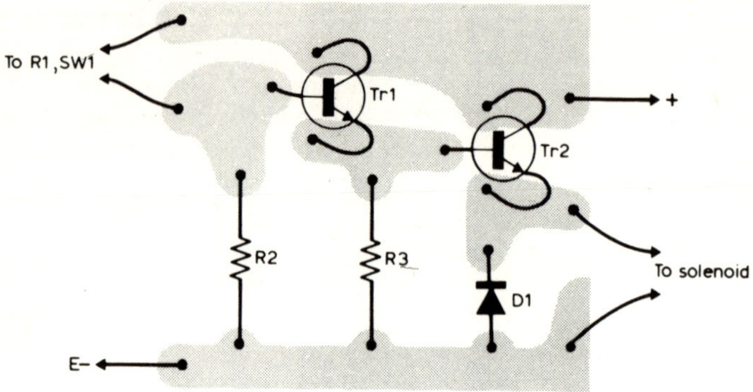

Fig. 9.2 PCB layout for solenoid/relay control circuit of Fig.2.2

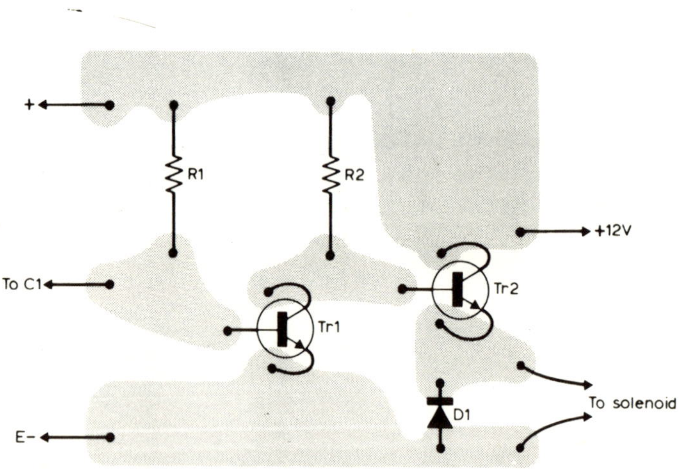

Fig. 9.3 PCB layout for latching solenoid circuit of Fig. 2.6

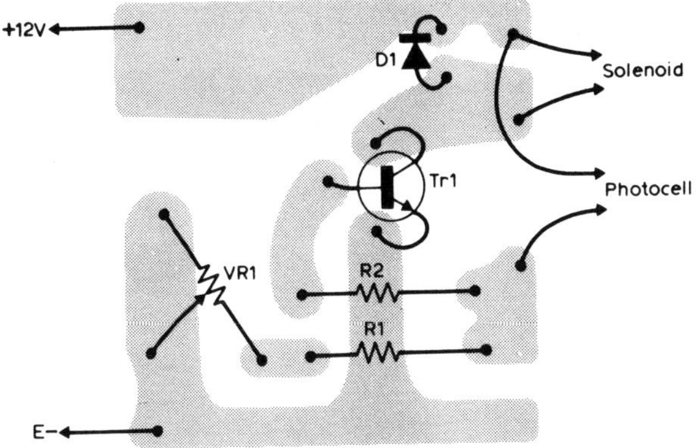

Fig. 9.4 PCB layout for light beam operated circuit of Fig. 2.7

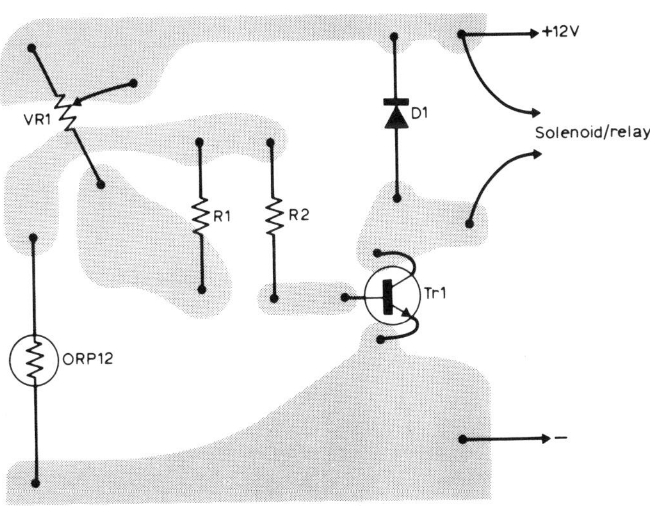

Fig. 9.5 PCB layout for light beam circuit of Fig. 2.8

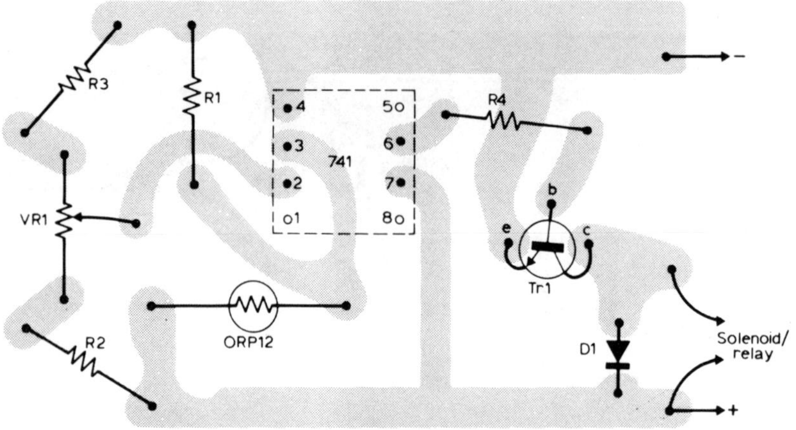

Fig. 9.6 PCB layout for op-amp circuit of Fig. 2.9

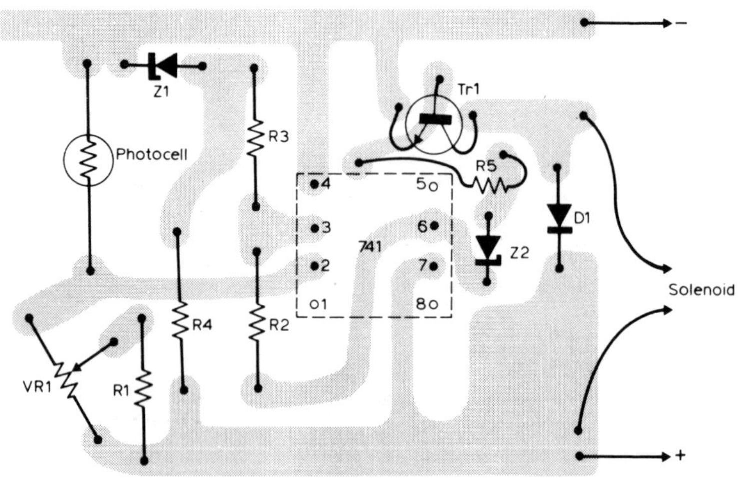

Fig. 9.7 PCB layout for op-amp circuit of Fig. 2.11

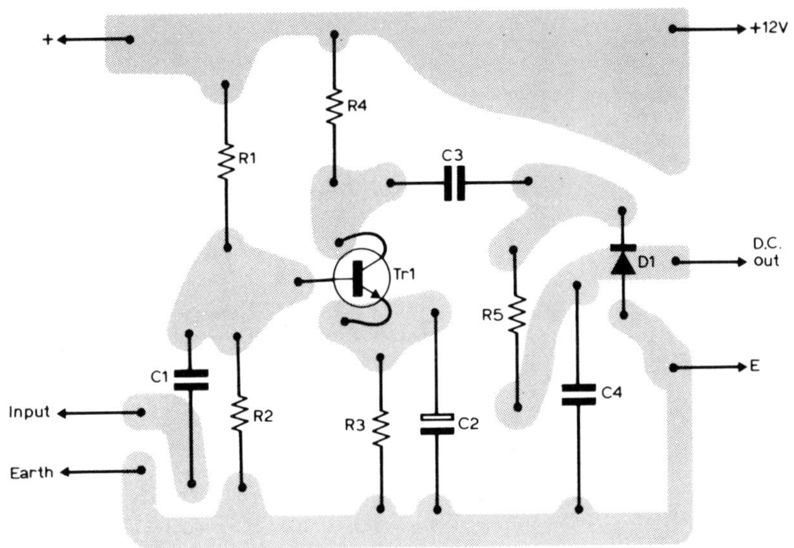

Fig. 9.8 PCB layout for a.c.-to-d.c. converter of Fig. 2.12

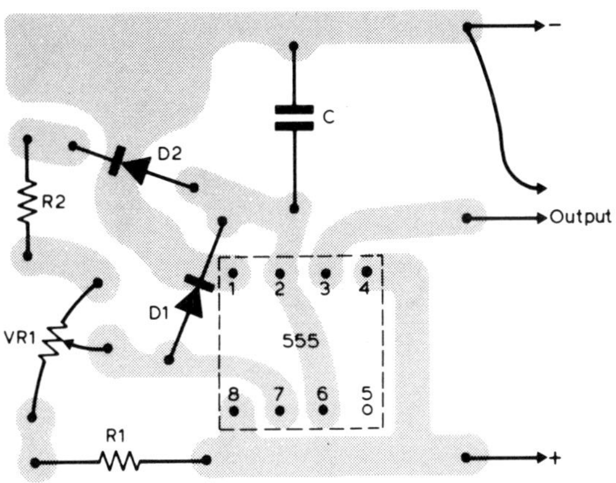

Fig. 9.9 PCB layout for timer circuit of Fig. 3.3

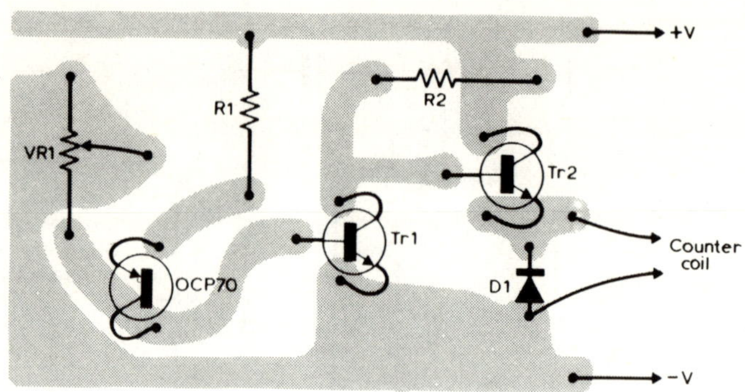

Fig. 9.10 PCB layout for light beam detector circuit of Fig. 5.3

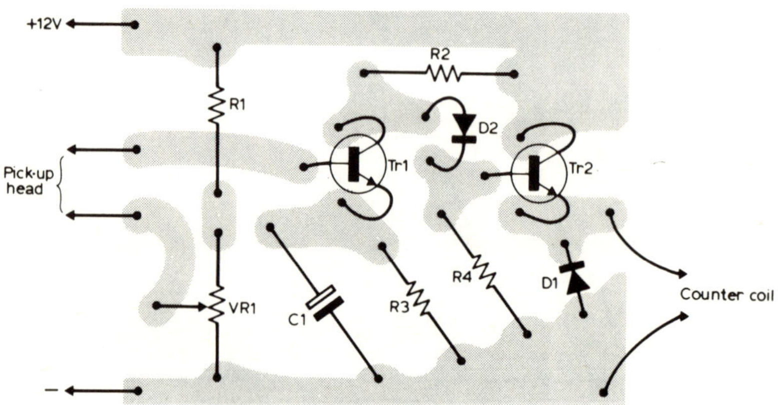

Fig. 9.11 PCB layout for magnetic detection circuit of Fig. 5.5

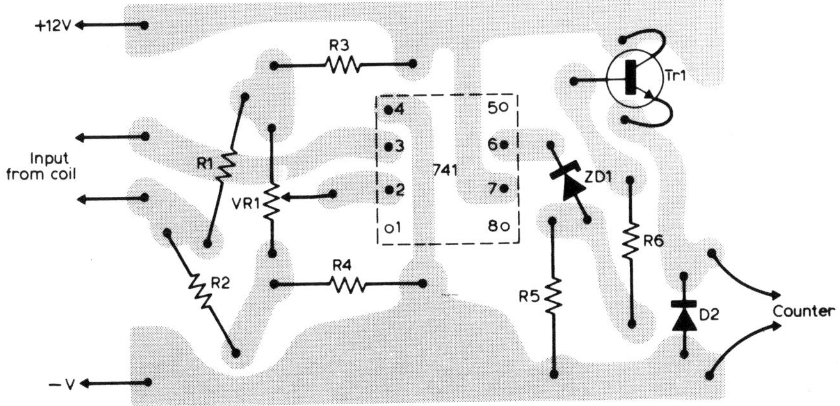

Fig. 9.12 PCB layout for magnetic detection circuit of Fig. 5.6

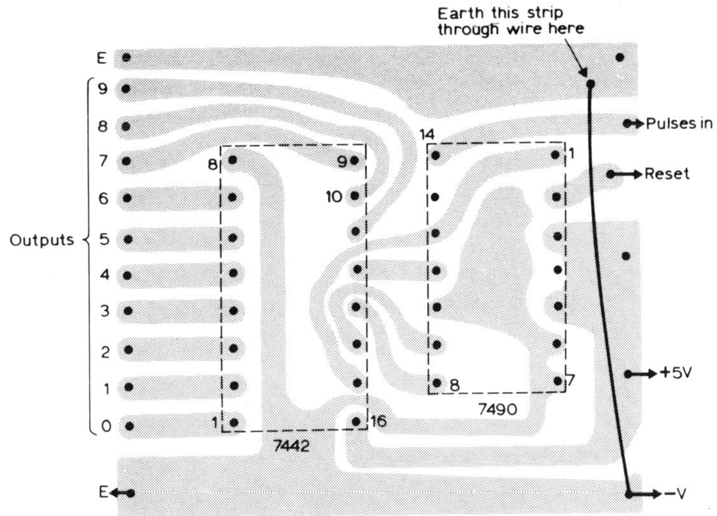

Fig. 9.13 PCB layout for pulse counting unit of Fig. 6.11

NOTE ON SUPPLIES

Electronic components are obtainable from a very large number of specialists, who advertise regularly in the magazines devoted to electronics. The components used in the circuits described in this book have been obtained from a wide variety of sources, all by mail-order. The following firms provided items which were used in the design of circuits:

DORAM ELECTRONICS Ltd.,
P.O. Box TR8,
Wellington Road Industrial Estate,
Wellington Bridge,
LEEDS LS2 2UF

For all RS COMPONENTS materials and integrated circuits mentioned in this book, and a very large range of electronic components. The catalogue alone is a good start, with many items which will be of interest to modellers.

BI-PAK,
P.O. Box 6,
Ware,
Herts.

Mainly suppliers of semiconductors at very low prices. Excellent source of integrated circuits.

BI-PRE-PAK Ltd.,
222/224 West Road,
WESTCLIFF-ON-SEA,
Essex. SS0 9DF

Supply a wide range of components at very low prices. Untested PAKS of transistors are ideal for experimental purposes; and the 2N697's used for testing various circuits in this book were from such a PAK.

ABBREVIATIONS

Because of the limited space on electronics circuit diagrams, abbreviations are frequently used. Some of the most common abbreviations which will be found in this book are shown below.

V volt(s)	potential difference	electrical units
I amps	electric current	
Ω ohms	resistance	
F farads	capacitance	
m milli	= 1/1000	multiples and submultiples
μ micro	= 1/1000000	
p pico	= 1/1000000000000	
n nano	= 1/1000000000	
k kilo	= 1000	
M mega	= 1000000	

NOTE: 1μF = 1000nF; 1nF = 1000pF. When the multiples k and M are used with Ω for resistance, the Ω sign is often omitted, so that a resistance value may be written as 47k, 2·2M etc. Recently, to avoid confusion which may be caused if the decimal point does not reproduce well in diagrams, the multiplier is placed where the decimal point would go, with R used where the resistance is less than 1000 ohms.

Examples: 2·2k = 2k2; 470Ω = 470R; 1·2M = 1M2; 5.6Ω = 5R6